Lecture Notes in Computer Science 13894

Founding Editors

Gerhard Goos
Juris Hartmanis

Editorial Board Members

The series Lecture Notes in Computer Science (LNCS), including its subseries Lecture Notes in Artificial Intelligence (LNAI) and Lecture Notes in Bioinformatics (LNBI), has established itself as a medium for the publication of new developments in computer science and information technology research, teaching, and education.

LNCS enjoys close cooperation with the computer science R & D community, the series counts many renowned academics among its volume editors and paper authors, and collaborates with prestigious societies. Its mission is to serve this international community by providing an invaluable service, mainly focused on the publication of conference and workshop proceedings and postproceedings. LNCS commenced publication in 1973.

Megan Dewar · Paweł Prałat ·
Przemysław Szufel · François Théberge ·
Małgorzata Wrzosek

Editors

Algorithms and Models for the Web Graph

18th International Workshop, WAW 2023
Toronto, ON, Canada, May 23–26, 2023
Proceedings

 Springer

Editors
Megan Dewar
Tutte Institute for Mathematics
and Computing
Ottawa, ON, Canada

Paweł Prałat
Toronto Metropolitan University
Toronto, ON, Canada

Przemysław Szufel
SGH Warsaw School of Economics
Warsaw, Poland

François Théberge
Tutte Institute for Mathematics
and Computing
Ottawa, ON, Canada

Małgorzata Wrzosek
SGH Warsaw School of Economics
Warsaw, Poland

ISSN 0302-9743 ISSN 1611-3349 (electronic)
Lecture Notes in Computer Science
ISBN 978-3-031-32295-2 ISBN 978-3-031-32296-9 (eBook)
https://doi.org/10.1007/978-3-031-32296-9

This Springer imprint is published by the registered company Springer Nature Switzerland AG
The registered company address is: Gewerbestrasse 11, 6330 Cham, Switzerland

Preface

The 18th Workshop on Algorithms and Models for the Web Graph (WAW 2023) was held at the Fields Institute for Research in Mathematical Sciences, Toronto, Canada (May 23–26, 2023). This is an annual meeting, which is traditionally co-located with another, related, conference. WAW 2023 was co-located with the Workshop on Modelling and Mining Complex Networks as Hypergraphs, which was held at the Toronto Metropolitan University, Toronto, Canada (May 15–19, 2023). Co-location of the two workshops provides opportunities for researchers in two different but interrelated areas to interact and to exchange research ideas. We do hope that both events were effective venues for the dissemination of new results and for fostering research collaboration.

The World Wide Web has become part of our everyday life, and information retrieval and data mining on the Web are now of enormous practical interest. The algorithms supporting these activities combine the view of the Web as a text repository and as a graph, induced in various ways by links among pages, hosts and users. The aim of the workshop was to further the understanding of graphs that arise from the Web and various user activities on the Web, and stimulate the development of high-performance algorithms and applications that exploit these graphs. The workshop gathered together researchers who are working on graph-theoretic and algorithmic aspects of related complex networks, including social networks, citation networks, biological networks, molecular networks, and other networks arising from the Internet.

This volume contains the papers accepted and presented during the workshop. Each submission was carefully reviewed by the members of the Programme Committee. Papers were submitted and reviewed using the EasyChair online system. The committee members decided to accept 12 papers.

May 2023

Megan Dewar
Paweł Prałat
Przemysław Szufel
François Théberge
Małgorzata Wrzosek

Organization

General Chairs

Andrei Z. Broder   Google Research, USA
Fan Chung Graham   University of California San Diego, USA

Organizing Committee

Megan Dewar   Tutte Institute for Mathematics and Computing,
                                        Canada
Paweł Prałat   Toronto Metropolitan University, Canada
Przemysław Szufel   SGH Warsaw School of Economics, Poland
François Théberge   Tutte Institute for Mathematics and Computing,
                                        Canada
Małgorzata Wrzosek   SGH Warsaw School of Economics, Poland

Sponsoring Institutions

Fields Institute for Research in Mathematical Sciences
NAWA – The Polish National Agency for Academic Exchange
SGH Warsaw School of Economics
Toronto Metropolitan University
Tutte Institute for Mathematics and Computing
Google

Program Committee

Konstantin Avratchenkov   Inria, France
Mindaugas Bloznelis   Vilnius University, Lithuania
Paolo Boldi   University of Milan, Italy
Anthony Bonato   Toronto Metropolitan University, Canada
Ulrik Brandes   ETH Zürich, Switzerland
Fan Chung Graham   UC San Diego, USA
Collin Cooper   King's College London, UK
Andrzej Dudek   Western Michigan University, USA

Contents

Correcting for Granularity Bias in Modularity-Based Community
Detection Methods 1
    Martijn Gösgens, Remco van der Hofstad, and Nelly Litvak

The Emergence of a Giant Component in One-Dimensional
Inhomogeneous Networks with Long-Range Effects 19
    Peter Gracar, Lukas Lüchtrath, and Christian Mönch

Unsupervised Framework for Evaluating Structural Node Embeddings
of Graphs 36
    Ashkan Dehghan, Kinga Siuta, Agata Skorupka, Andrei Betlen,
    David Miller, Bogumił Kamiński, and Paweł Prałat

Modularity Based Community Detection in Hypergraphs 52
    Bogumił Kamiński, Paweł Misiorek, Paweł Prałat, and François Théberge

Establishing Herd Immunity is Hard Even in Simple Geometric Networks 68
    Michal Dvořák, Dušan Knop, and Šimon Schierreich

Multilayer Hypergraph Clustering Using the Aggregate Similarity Matrix 83
    Kalle Alaluusua, Konstantin Avrachenkov, B. R. Vinay Kumar,
    and Lasse Leskelä

The Myth of the Robust-Yet-Fragile Nature of Scale-Free Networks:
An Empirical Analysis 99
    Rouzbeh Hasheminezhad, August Bøgh Rønberg, and Ulrik Brandes

A Random Graph Model for Clustering Graphs 112
    Fan Chung and Nicholas Sieger

Topological Analysis of Temporal Hypergraphs 127
    Audun Myers, Cliff Joslyn, Bill Kay, Emilie Purvine, Gregory Roek,
    and Madelyn Shapiro

PageRank Nibble on the Sparse Directed Stochastic Block Model 147
    Sayan Banerjee, Prabhanka Deka, and Mariana Olvera-Cravioto

A Simple Model of Influence 164
    Colin Cooper, Nan Kang, and Tomasz Radzik

x   Contents

The Iterated Local Transitivity Model for Tournaments 179
  Anthony Bonato and Ketan Chaudhary

Author Index 193

Correcting for Granularity Bias in Modularity-Based Community Detection Methods

Martijn Gösgens[1]([✉])[iD], Remco van der Hofstad[1][iD], and Nelly Litvak[1,2][iD]

[1] Eindhoven University of Technology, Eindhoven, Netherlands
research@martijngosgens.nl
[2] University of Twente, Enschede, Netherlands

Abstract. Maximizing modularity is currently the most widely-used
community detection method in applications. Modularity comes with a
parameter that indirectly controls the granularity of the resulting clustering. Moreover, one can choose this parameter in such a way that modularity maximization becomes equivalent to maximizing the likelihood of a
stochastic block model. Thus, this method is statistically justified, while
at the same time, it is known to have a bias towards fine-grained clusterings. In this work, we introduce a heuristic to correct for this bias. This
heuristic is based on prior work where modularity is described in geometric terms. This has led to a broad generalization of modularity-based
community detection methods, and the heuristic presented in this paper
applies to each of them. We justify the heuristic by describing a relation
between several distances that we observe to hold in many instances. We
prove that, assuming the validity of this relation, our heuristic leads to
a clustering of the same granularity as the ground-truth clustering. We
compare our heuristic to likelihood-based community detection methods
on several synthetic graphs and show that our method indeed results in
clusterings with granularity closer to the granularity of the ground-truth
clustering. Moreover, our heuristic often outperforms likelihood maximization in terms of similarity to the ground-truth clustering.

Keywords: Community detection · Clustering · Modularity ·
Likelihood maximization

1 Introduction

One of the most widely-used community detection methods is the maximization of *modularity* [16,22]. Modularity measures the fraction of edges that lie
inside communities minus the expectation of this fraction in a *null model*, a
random graph without community structure. Two popular null models are the

Supported by the Netherlands Organisation for Scientific Research (NWO) through
the Gravitation NETWORKS grant no. 024.002.003.

Erdős-Rényi model and the Configuration Model[1], and we will refer to the corresponding modularity measures as ER- and CM-modularity, respectively.

Modularity comes with a *resolution parameter*, which controls the granularity of the resulting clustering [3,11,12]. However, it is not obvious how to choose this parameter so that the obtained clustering has the right granularity [1,18]. It has been proven that with a particular value of the resolution parameter, maximizing ER-modularity is equivalent to maximizing the likelihood of a Planted Partition Model (PPM), while maximizing CM-modularity is equivalent [15] to maximizing the likelihood of a Degree-Corrected PPM (DCPPM). This equivalence justifies the corresponding specific choice of the resolution parameter, that turns modularity maximization into likelihood maximization. However, it has been observed [17,24] that the resulting Maximum-Likelihood Estimator (MLE) is biased towards fine-grained clusterings. That is, the method typically finds more and smaller communities than are present in the ground-truth clustering. This behavior even occurs when maximizing the PPM-likelihood on PPM graphs, as we will demonstrate in Sect. 5.

In this paper, we introduce a heuristic to correct for this *granularity bias*. To do so, we build on prior work [4], where we proved that maximizing modularity is equivalent to minimizing the distance between a clustering and some *modularity vector* on a hypersphere. This equivalence leads to a geometric framework in which modularity-based methods can be studied. Additionally, this paves the way to a broad class of new community detection methods that often outperform modularity-based methods. Our proposed heuristic can be applied to any modularity-based method. Additionally, it can be applied to any type of pairwise similarity data between the vertices. For example, applying our heuristic to Jaccard similarities between vertex-neighborhoods results in a method that outperforms likelihood-based methods in several random graphs, as we demonstrate in Sect. 5. Our Python code is available on GitHub[2].

Notation. In this paper, we consider graphs consisting of n vertices and we let $N = \binom{n}{2}$ denote the number of vertex-pairs. Let G be a graph with vertex-set $[n] = \{1, 2, \ldots, n\}$ and let the degree of vertex i be given by $d_i^{(G)}$. We denote the neighborhood of a vertex i by $N(i)$. We consider every vertex to be part of its own neighborhood. That is, $i \in N(i)$ and $|N(i)| = d_i^{(G)} + 1$ for any $i \in [n]$. The number of edges in the graph is given by $m_G = \frac{1}{2} \sum_{i \in [n]} d_i^{(G)}$. Let C be a clustering (i.e., a partition of $[n]$). We denote the number of intra-cluster pairs of C by m_C.

2   Hyperspherical Geometry

In this section, we introduce the hyperspherical geometry that we use in the remainder of this paper. A more elaborate description is given in [4].

---

[1] Or, equivalently, the Chung-Lu model.
[2] https://github.com/MartijnGosgens/hyperspherical_community_detection.

We represent clusterings as high-dimensional vectors. Let C be a clustering, then we define the *clustering vector* $\boldsymbol{b}(C)$ as the binary vector indexed by vertex-pairs, with elements ij, where $i, j \in [n]$, given by

$$\boldsymbol{b}(C)_{ij} = \begin{cases} 1, & \text{if } i \text{ and } j \text{ are in the same cluster, and} \\ -1, & \text{if } i \text{ and } j \text{ are in different clusters.} \end{cases} \tag{1}$$

Note that $\boldsymbol{b}(C) \in \mathbb{R}^N$. If a clustering consists of a single cluster of size n, then the corresponding clustering vector is the all-one vector $\boldsymbol{1}$. If a clustering consists of n clusters of size 1 each, then the corresponding clustering vector is $-\boldsymbol{1}$.

For each clustering C, the vector $\boldsymbol{b}(C)$ has Euclidean length $\sqrt{N}$, so that all clustering vectors lie on a hypersphere with radius $\sqrt{N}$, centered at the all-zero vector $\boldsymbol{0}$. This suggests a natural *hyperspherical geometry* induced by the *angular distance*. For two vectors $\boldsymbol{x}, \boldsymbol{y} \in \mathbb{R}^N \setminus \{\boldsymbol{0}\}$, the angular distance is given by

$$d_a(\boldsymbol{x}, \boldsymbol{y}) = \arccos\left(\frac{\langle \boldsymbol{x}, \boldsymbol{y} \rangle}{\|\boldsymbol{x}\| \cdot \|\boldsymbol{y}\|}\right), \tag{2}$$

where $\langle \cdot, \cdot \rangle$ denotes the standard inner product between two vertices, while $\|\boldsymbol{x}\| = \sqrt{\langle \boldsymbol{x}, \boldsymbol{x} \rangle}$ denotes the Euclidean norm of a vector. Since our geometry is induced by the angular distance, the lengths of vectors are of no importance and each of the definitions in the remainder of this section is invariant to vector length.

Equivalence to Modularity Maximization. In [4], it is proven that the angular distance between a clustering vector and a so-called *modularity vector* is a monotone transformation of the well-known modularity function. This implies that maximizing modularity is equivalent to finding the clustering vector that minimizes the angular distance to this modularity vector. For ER-modularity with resolution parameter γ, this modularity vector is given by

$$\boldsymbol{q}_{\text{ER}}^{(\text{Mod})}(G, \gamma) = \boldsymbol{1} + \boldsymbol{e}(G) - 2\gamma \frac{m_G}{N}\boldsymbol{1}, \tag{3}$$

where $\boldsymbol{e}(G)$ is the *edge vector*, with $\boldsymbol{e}(G)_{ij} = 1$ if i and j share an edge and $\boldsymbol{e}(G)_{ij} = -1$ otherwise. For CM-modularity, the corresponding modularity vector is given by

$$\boldsymbol{q}_{\text{CM}}^{(\text{Mod})}(G, \gamma) = \boldsymbol{1} + \boldsymbol{e}(G) - 2\gamma \cdot \boldsymbol{d}(G), \tag{4}$$

where $\boldsymbol{d}(G)$ is a *degree-correction* vector, given by $\boldsymbol{d}(G)_{ij} = \frac{1}{2m_G}d_i^{(G)}d_j^{(G)}$. However, we can replace this modularity vector by any other vector to obtain a different clustering method. We define the resulting class of clustering methods as *projection methods*, since we are *projecting* a vertex to the set of clustering vectors. We refer to the point that is being projected as the *query vector*.

Definition 1. *A projection method is a clustering method where a candidate clustering is obtained by minimizing the angular distance to some query vector $\boldsymbol{q}$. That is, we find the clustering C that minimizes $d_a(\boldsymbol{q}, \boldsymbol{b}(C))$.*

The Louvain Projection. From this viewpoint, any algorithm that maximizes modularity can equivalently be viewed as an algorithm that projects a query vector on to the set of clustering vectors. The Louvain algorithm is one of the most widely-used modularity-maximization algorithm [2]. Moreover, this algorithm can easily be modified to minimize distances to other query vectors. Throughout this paper, we will use the Louvain algorithm to find clustering vectors that approximately minimize the distance to query vectors. Let $\mathcal{L}(q)$ denote the clustering vector that results from applying the Louvain algorithm to a query vector q. More details about how we modify the Louvain algorithm to minimize distances to different query vectors can be found in [4].

Latitude and Granularity. Another nice feature of the hyperspherical geometry is that the granularity of a clustering can be interpreted in geometric terms. We define the *latitude* $\ell(x)$ of a vector x as its angular distance to -1, i.e., $\ell(x) = d_a(x, -1)$. For clustering vectors, this latitude is given by

$$\ell(b(C)) = \arccos\left(1 - 2\frac{m_C}{N}\right), \tag{5}$$

which is a monotone transformation of m_C/N, the fraction of vertex-pairs that are intra-cluster pairs. This fraction (otherwise known as the Simpson's index) is a measure of granularity that is often used in ecology [7].

Parallels and Meridians. In globe terminology, a *parallel* is a set of a constant latitude. In a three-dimensional globe, a parallel corresponds to a circle, while in our case it corresponds to a high-dimensional surface. For example, the *equator* is the parallel with latitude $\pi/2$ and it consists of all vertices x that satisfy $\langle x, 1 \rangle = 0$. We define the *parallel projection* $\mathcal{P}_\lambda(x)$ as the projection of x to the parallel with latitude λ. For any vector x that is not a multiple of 1 and any $\lambda \in [0, \pi]$, this projection is given by

$$\mathcal{P}_\lambda(x) = \sin(\lambda)\frac{\sqrt{N}}{\|x - \frac{\langle x, 1 \rangle}{N}1\|}\left(x - \frac{\langle x, 1 \rangle}{N}1\right) - \cos(\lambda)1. \tag{6}$$

Similarly, a *meridian* is defined in globe terminology as a set of constant longitude. In our high-dimensional sphere, longitude cannot be defined as a scalar value. However, for any vector x that is not a multiple of 1, we can define its meridian as the set $\{\mathcal{P}_\lambda(x) : \lambda \in (0, \pi)\}$. The modularity vector corresponding to ER-modularity lies on the meridian of the vector $e(G)$, and there is a relation between the resolution parameter and the latitude of the modularity vector [4].

Validation Measures. For clustering vectors, the angular distance is a monotone transformation [4] of the Rand index, which is a well-known similarity measure between clusterings [21]. Such similarity measures are used to measure the performance of a clustering algorithm in settings where a ground-truth clustering is available. Specifically, if T is the ground-truth clustering while C is a candidate clustering, then the similarity between C and T can be used to measure the performance of the clustering method. There exist many such clustering similarity measures [6]. In this paper, we will use the *correlation distance* to measure similarity between clusterings. The correlation distance between clusterings C_1 and C_2 is the arccosine of the Pearson correlation coefficient between the clustering vectors $b(C_1)$ and $b(C_2)$. We choose this measure because it satisfies many desirable properties [5,6] and because it has a nice interpretation in terms of the hyperspherical geometry. The correlation distance is given by

$$d_{\text{CC}}(\boldsymbol{x}, \boldsymbol{y}) = \arccos\left(\frac{\cos d_a(\boldsymbol{x}, \boldsymbol{y}) - \cos \ell(\boldsymbol{x}) \cos \ell(\boldsymbol{y})}{\sin \ell(\boldsymbol{x}) \sin \ell(\boldsymbol{y})}\right), \tag{7}$$

and can be interpreted as the angle that the meridians of $\boldsymbol{x}$ and $\boldsymbol{y}$ make when they intersect at $-\mathbf{1}$. In this paper, we will use $d_{\text{CC}}(b(C), b(T))$ as a performance measure of a clustering method that has produced C, *lower* values indicating better performance.

3   The Heuristic

We assume that we are given some pair-wise similarity data of the vertices, represented by a vector $\boldsymbol{x} \in \mathbb{R}^N$ indexed by vertex-pairs, which we call the *input vector*. For example, we could take $\boldsymbol{x} = e(G)$. Our objective is to find a latitude $\lambda^* \in [0, \pi]$ so that the vector $\boldsymbol{q}^*(\boldsymbol{x}) = \mathcal{P}_{\lambda^*}(\boldsymbol{x})$ makes for a good query vector. Ideally, we want $\mathcal{L}(\boldsymbol{q}^*)$ to be close to $b(T)$. Since we measure the proximity between clusterings by the correlation distance, we want to choose λ^* such that $d_{\text{CC}}(\mathcal{L}(\boldsymbol{q}^*), b(T))$ is minimized. However, this is problematic because there is no method to analytically derive the output of the Louvain algorithm. Therefore, we cannot obtain a mathematical relation between the latitude λ of query $\boldsymbol{q}$ and performance measure $d_{\text{CC}}(\mathcal{L}(\boldsymbol{q}), b(T))$.

A more feasible goal is to choose the query latitude such that the candidate clustering has a similar latitude as the ground-truth clustering. That is, we want to choose λ^* such that $\boldsymbol{q}^* = \mathcal{P}_{\lambda^*}(\boldsymbol{x})$ achieves $\ell(\mathcal{L}(\boldsymbol{q}^*)) \approx \ell(b(T))$. Since the latitude of a clustering is a measure of its granularity, this means that the candidate clustering will have a similar granularity as the ground-truth clustering. For this simpler goal, it turns out that we can obtain some guidelines of how to choose λ^*, from numerical experiments. Moreover, in our experiments, we observe that this choice of λ^* also leads to a near-minimal correlation distance between $\mathcal{L}(\boldsymbol{q}^*)$ and $b(T)$, so that the objective of finding a clustering close to the ground truth is also approximately achieved. In this section, we present our proposed query latitude, and in Sect. 4, we show how we derive this heuristic from empirical observations.

Heuristic Latitude. Our proposed query latitude choice is given by

$$\lambda^*(\lambda_T, \theta) = \arccos\left(\frac{\cos \lambda_T \cos \theta}{1 + \sin \lambda_T \sin \theta}\right), \tag{8}$$

where $\lambda_T = \ell(\boldsymbol{b}(T))$ is the latitude of the ground truth clustering and $\theta = d_{\text{CC}}(\boldsymbol{x}, \boldsymbol{b}(T))$ is the correlation distance between the ground truth and the input vector. The query vector that our heuristic prescribes is thus given by

$$q^*(\boldsymbol{x}) = \mathcal{P}_{\lambda^*(\lambda_T, \theta)}(\boldsymbol{x}). \tag{9}$$

In summary, our geometrically-inspired community detection method clusters the vertices by evaluating the expression $\mathcal{L}(\boldsymbol{q}^*(\boldsymbol{x}))$. That is, we propose to find communities by first mapping $\boldsymbol{x}$ to the query vector $\boldsymbol{q}^* = \mathcal{P}_{\lambda^*}(\boldsymbol{x})$, and then computing the projection $\mathcal{L}(\boldsymbol{q}^*)$ with the Louvain algorithm.

Requiring Partial Knowledge of Ground Truth. Our heuristic assumes that the latitude λ_T of the ground truth clustering $\boldsymbol{b}(T)$, as well as the correlation distance θ between $\boldsymbol{b}(T)$ and input vector $\boldsymbol{x}$ are known or that approximations of these are available. At first, it may seem that requiring such knowledge about $\boldsymbol{b}(T)$ defeats the purpose of community detection. However, assuming this knowledge is not uncommon in other community detection methods, such as likelihood-based methods [15,19]. For instance, likelihood maximization using the Planted Partition Model, requires knowledge of the intra- and inter-community edge densities. These densities can be combined with the total density of the graph to compute the number of intra-community pairs m_T, which is related to λ_T by (5). Then, we can compute the number of intra- and inter-community edges, from which $d_{\text{CC}}(\boldsymbol{x}, \boldsymbol{b}(T))$ can be computed for $\boldsymbol{x} = e(G)$. Overall, our method requires the *same* information as maximum-likelihood methods based on the Planted Partition Model or its degree-corrected variant. This suggests that the required knowledge about λ_T and θ is comparable to existing methods, and therefore is acceptable in practice.

When applying likelihood-based methods in practice, the issue of requiring estimates of the ground-truth clustering is usually overcome by applying the method iteratively and updating the parameters after each iteration based on the obtained candidate clustering. A comparable approach could be taken for our heuristic, as we will briefly discuss in Sect. 4.

Input Vectors. When choosing $\boldsymbol{x} = e(G)$, the query vector that results from our heuristic is equivalent to ER-modularity for some resolution parameter $\gamma(\lambda^*)$. The Maximum Likelihood Estimator (MLE) for PPM also corresponds to ER-modularity [15]. In Sect. 5, we will see that our heuristic with $\boldsymbol{x} = e(G)$ has a similar performance as PPM-MLE when applied to PPM graphs, while being more accurate in terms of the clustering latitude. In this sense, our heuristic can be seen as an improvement to PPM-MLE that is *corrected for granularity bias*.

Moreover, and even more importantly, the strength of our heuristic is that it works for a wide class of input vectors. Any pair-wise similarity measure could be used for this purpose. For example, instead of using the input vector $e(G)$, we could use the query vector corresponding to the MLE of the Degree-Corrected PPM. The resulting query vector does not correspond to a known modularity function. In Sect. 5, we will see that, besides resulting in clusterings with more accurate latitudes, this also results in a slight improvement of the performance.

Besides the information on edges and degrees of vertices, as in the previous examples, x can also include other information. For example, in prior work [4], we have demonstrated that query vectors based on common neighbors perform well on some networks. We call the corresponding input vector a *wedge vector* $w(G)$, where each element $w(G)_{ij}$ equals to the number of common neighbors of i and j, or, equivalently, the number of wedges that have i and j as their endpoints.

Another class of input vectors could be obtained by measuring the similarity between the neighborhoods of vertices. For this purpose, it is convenient to consider each vertex as part of its own neighborhoods (i.e., $i \in N(i)$ for every i), so that an edge between i and j already implies an overlap of at least 2. We call the corresponding input vector a *Jaccard vector*, where each element $j(G)_{ij}$ is the *Jaccard* similarity [8] between the neighborhoods of i and j:

$$j(G)_{ij} = \frac{|N(i) \cap N(j)|}{|N(i) \cup N(j)|}, \quad \text{for all pairs } ij.$$

In Sect. 5, we will see that this input vector, combined with the heuristic, performs well in practice. Instead of the Jaccard index, any function that measures similarity between sets can be used. The advantage of the Jaccard index is that it equals zero when neighborhoods do not overlap, which is beneficial for the running time of the Louvain algorithm.

Besides the input vectors mentioned above, any linear combination of them could also be used. This gives infinitely many input vectors to choose from, and it is difficult to predict which performs best on a particular graph. The correlation distance $d_{CC}(x, b(T))$ can be used to indicate which input is most suitable for a particular ground-truth clustering.

4 Derivation of the Heuristic

Ideally, we want to choose the query vector q such that $\mathcal{L}(q)$ is as close as possible to $b(T)$. Since there is no method to predict $\mathcal{L}(q)$ analytically, we have conducted a large number of experiments (not presented here for brevity), where we have varied $\ell(q)$ and observed the properties of the desired query vector q^*. We observed that for q^*, the four distances $d_a(q^*, b(T))$, $d_{CC}(q^*, b(T))$, $d_a(q^*, \mathcal{L}(q^*))$ and $d_{CC}(q^*, \mathcal{L}(q^*))$, are approximately equal to each other. When we take query vectors on a given meridian, i.e. $q(\lambda) = \mathcal{P}_\lambda(x)$, and consider these four distances as functions of the query latitude λ, then we see that the intersections of the curves $\lambda \mapsto d_a(q(\lambda), b(T))$ with $\lambda \mapsto d_{CC}(q(\lambda), b(T))$ and $\lambda \mapsto d_a(q(\lambda), \mathcal{L}(q(\lambda)))$ with $\lambda \mapsto d_{CC}(q(\lambda), \mathcal{L}(q(\lambda)))$ occur at approximately the same value of λ. This behavior would be trivial if $\mathcal{L}(q) = b(T)$, i.e., if the Louvain algorithm *exactly* recovered the ground-truth clustering. Interestingly, this behavior persists when the best-performing query vector results in a clustering with considerable dissimilarity to the ground truth. We have not been able to theoretically explain this behavior, but it will prove very useful in deriving our heuristic.

In addition, we use one property of the Louvain algorithm that is easy to prove [4]: the inequality $d_a(q, \mathcal{L}(q)) \leq \ell(q)$ holds for any query vector q. This is because the Louvain algorithm finds candidate clustering C by greedily decreasing $d_a(q, b(C))$. Since Louvain initializes with n clusters of size 1, with corresponding clustering vector equal to -1, the objective function initially equals $d_a(q, -1) = \ell(q)$, and it can only decrease in subsequent iterations. We combine this provable constraint with the behaviour described above, in the next empirical observation:

Observation 1. *Whenever* $d_{\mathrm{cc}}(q, b(T)) = d_a(q, b(T)) = \theta$ *for some* $\theta \in (0, \ell(q))$, *then*

$$d_a(q, \mathcal{L}(q)) \approx \theta, \qquad and \qquad d_{\mathrm{cc}}(q, \mathcal{L}(q)) \approx \theta.$$

We illustrate Observation 1 for two input vectors (or, rather, their meridians). The first input vector is $e(G)$, which is equivalent to the ER-modularity. The second input vector $r(T)$ is the ground-truth clustering vector plus some random Gaussian noise: $r(T) = b(T) + r$, where each entry of r is independently drawn from the normal distribution with mean zero and standard deviation 1.5 (chosen such that $d_{\mathrm{cc}}(r(T), b(T)) \approx d_{\mathrm{cc}}(e(G), b(T))$).

In the experiment illustrated in Figs. 1a and 1b, we have computed the intersection point λ^* of $d_a(q(\lambda), b(T))$ with $d_{\mathrm{cc}}(q(\lambda), b(T))$, and denoted it in the figure with a cross. The figure shows the lines $d_a(q(\lambda), \mathcal{L}(q(\lambda)))$ and $d_{\mathrm{cc}}(q(\lambda), \mathcal{L}(q(\lambda)))$. We see that they indeed intersect approximately at $\lambda = \lambda^*$ as claimed by Observation 1. Figures 1c and 1d show that the latitude of the candidate clustering is approximately equal to the latitude of the ground-truth when the latitude of the query is $\lambda = \lambda^*$. This consequence of Observation 1 will be proved in Theorem 1. Finally, Figs. 1e and 1f show that the best performance is also achieved around latitude λ^*.

To further validate Observation 1, we conduct the following experiment: for a clustering consisting of 10 clusters of size 10 each, we take input vector $x = r(T)$ with standard deviations $\sigma \in [0.5, 2]$. Then we take the query vector $q^*(r(T))$ on the meridian of $r(T)$ such that $d_a(q^*, b(T)) = d_{\mathrm{cc}}(q^*, b(T))$ and evaluate $d_a(q^*, \mathcal{L}(q^*))/d_a(q^*, b(T))$ and $d_{\mathrm{cc}}(q^*, \mathcal{L}(q^*))/d_{\mathrm{cc}}(q^*, b(T))$. We do this 50 times for each standard deviation. The results in the form of box-plots are presented in Fig. 2. Values close to one indicate that Observation 1 holds with high accuracy. We see that for standard deviations up to $\sigma = 1$ (corresponding to $d_{\mathrm{cc}}(r(T), b(T)) \approx \pi/3$), the approximation holds almost exactly, while for larger standard deviations, the error becomes larger.

We now prove that the query vector given in (9) with latitude given by (8) indeed satisfies the conditions of Observation 1:

Lemma 1. *Suppose* $\lambda_T = \ell(b(T))$ *and* $\theta = d_{\mathrm{cc}}(x, b(T)) < \pi/2$. *The query vector* $q^* = \mathcal{P}_{\lambda^*}(x)$, *with* $\lambda^* = \lambda^*(\lambda_T, \theta)$ *given by* (8) *satisfies* $d_a(q^*, b(T)) = \theta$.

Proof. We compute the sine of λ^* as $\sin \lambda^* = \sqrt{1 - \cos^2 \lambda^*}$, which gives

$$\sin \lambda^* = \frac{\sin \lambda_T + \sin \theta}{1 + \sin \lambda_T \sin \theta}. \tag{10}$$

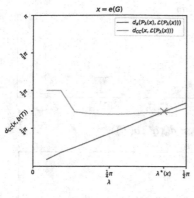

(a) Validating Observation 1 for $x = e(G)$.

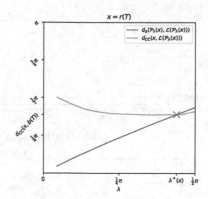

(b) Validating Observation 1 for $x = r(T)$.

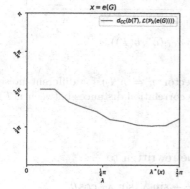

(c) Evaluating Theorem 1 empirically for $x = e(G)$.

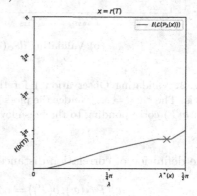

(d) Evaluating Theorem 1 empirically for $x = r(T)$.

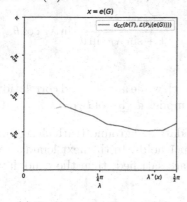

(e) Performance of $x = e(G)$.

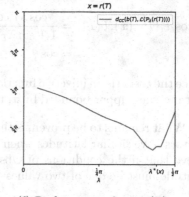

(f) Performance of $x = r(T)$.

Fig. 1. Evaluation of $\mathcal{L}(q^*(x))$ for $x \in \{e(G), r(T)\}$. The graphs are generated by the PPM with $n = 100$ vertices, 10 communities of size 10 each, and the average intra- and inter-community degree of 4. The crosses in Figs. 1a and 1b denote the intersection points claimed by Observation 1. Theorem 1 tells us that the lines of Figs. 1c and 1d should roughly intersect at the crosses. Figures 1e and 1f show the performance measures.

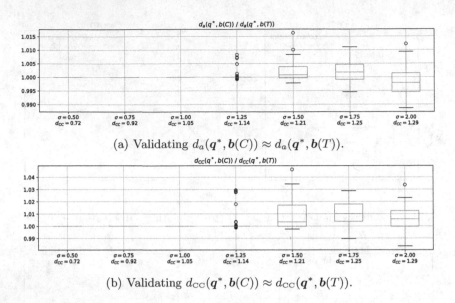

(a) Validating $d_a(\boldsymbol{q}^*, \boldsymbol{b}(C)) \approx d_a(\boldsymbol{q}^*, \boldsymbol{b}(T))$.

(b) Validating $d_{\mathrm{CC}}(\boldsymbol{q}^*, \boldsymbol{b}(C)) \approx d_{\mathrm{CC}}(\boldsymbol{q}^*, \boldsymbol{b}(T))$.

Fig. 2. Validating Observation 1 for the query vector $\boldsymbol{x} = \boldsymbol{r}(T)$ for different noise levels. The "$d_{\mathrm{CC}} = \dots$" under the plots refer to the correlation distances between $\boldsymbol{r}(T)$ and $\boldsymbol{b}(T)$ corresponding to the noise levels.

The definition of correlation distance (7) can be rewritten to

$$\cos d_a(\mathcal{P}_{\lambda^*}(\boldsymbol{x}), \boldsymbol{b}(T)) = \cos \lambda^* \cos \lambda_T + \sin \lambda^* \sin \lambda_T \cos \theta.$$

Now, if we substitute $\cos \lambda^*$ and $\sin \lambda^*$ from (8) and (10), then we get

$$\cos d_a(\mathcal{P}_{\lambda^*}(\boldsymbol{x}), \boldsymbol{b}(T)) = \frac{\cos \lambda_T \cos \theta}{1 + \sin \lambda_T \sin \theta} \cos \lambda_T + \frac{\sin \lambda_T + \sin \theta}{1 + \sin \lambda_T \sin \theta} \sin \lambda_T \cos \theta.$$
$$= \cos \theta.$$

Since the cosine is an injective function for $\theta \in [0, \pi]$, while angular and correlation distances are upper-bounded by π, this indeed implies $d_a(\boldsymbol{q}^*, \boldsymbol{b}(T)) = \theta$.   □

What remains to be proven, is that the candidate and ground-truth clustering vectors have similar latitudes when Observation 1 holds. In the next lemma, we prove that if the conditions of Observation 1 are satisfied, then the candidate latitude must be one of two values:

Lemma 2. *For a query vector $\boldsymbol{q}$ with $\lambda = \ell(\boldsymbol{q})$ and a clustering vector $\boldsymbol{b}(C)$ with $\ell(\boldsymbol{b}(C)) = \lambda_C$, if $d_a(\boldsymbol{q}, \boldsymbol{b}(C)) = d_{\mathrm{CC}}(\boldsymbol{q}, \boldsymbol{b}(C)) = \theta < \pi/2$, then*

$$\cos \lambda_C = \frac{\cos \lambda \cos \theta}{1 \pm \sin \lambda \sin \theta}.$$

Proof. The definition of correlation distance gives

$$\cos\theta = \frac{\cos d_a(\boldsymbol{q}, \boldsymbol{b}(C)) - \cos\ell(\boldsymbol{q})\cos\ell(\boldsymbol{b}(C))}{\sin\ell(\boldsymbol{q})\sin\ell(\boldsymbol{b}(C))} = \frac{\cos\theta - \cos\lambda\cos\lambda_C}{\sin\lambda\sin\lambda_C}.$$

We square both sides, substitute $\sin^2\lambda_C = 1 - \cos^2\lambda_C$ and rewrite the expression to obtain the following quadratic equation in $\cos\lambda_C$:

$$(\cos^2\lambda + \sin^2\lambda\cos^2\theta)\cos^2\lambda_C - 2\cos\theta\cos\lambda\cos\lambda_C + \cos^2\theta\cos^2\lambda = 0. \quad (11)$$

Note that the coefficient of the quadratic term can be rewritten to $1 - \sin^2\lambda\sin^2\theta$, which in turn can be factorized into $(1 - \sin\lambda\sin\theta)(1 + \sin\lambda\sin\theta)$. The solutions of (11) are given by

$$\cos\lambda_C = \frac{2\cos\lambda\cos\theta \pm \sqrt{4\cos^2\lambda\cos^2\theta - 4\cos^2\lambda\cos^2\theta(1 - \sin^2\lambda\sin^2\theta)}}{2(1 - \sin^2\lambda\sin^2\theta)}$$

$$= \cos\lambda\cos\theta\frac{1 \pm \sin\lambda\sin\theta}{1 - \sin^2\lambda\sin^2\theta} = \frac{\cos\lambda\cos\theta}{1 \mp \sin\lambda\sin\theta}.$$

$\square$

Lemma 2 gives two possible values for the candidate latitude that results from our heuristic. Specifically, for the query vector $\boldsymbol{q}^*$, assuming that Observation 1 holds, the Louvain algorithm will return $\boldsymbol{b}(C) = \mathcal{L}(\boldsymbol{q}^*)$ that satisfies one of the two expressions:

$$\ell(\mathcal{L}(\boldsymbol{q}^*)) \approx \arccos\left(\frac{\cos\lambda^*\cos\theta}{1 - \sin\lambda^*\sin\theta}\right), \text{ or } \ell(\mathcal{L}(\boldsymbol{q}^*)) \approx \arccos\left(\frac{\cos\lambda^*\cos\theta}{1 + \sin\lambda^*\sin\theta}\right).$$

We denote the first solution (with the minus) by λ_-, and denote the second solution by λ_+. Note that λ_+ is always closer to $\pi/2$ than λ_- since $\sin\theta > 0$. We have empirically observed that our heuristic tends to result in $\ell(\mathcal{L}(\boldsymbol{q}^*)) \approx \lambda_-$ and that $\ell(\mathcal{L}(\boldsymbol{q}^*)) \approx \lambda_+$ only occurs whenever both solutions are close (i.e., $\lambda_- \approx \lambda_+$). We illustrate this by the following experiment presented in Fig. 3: for clusterings of 100 vertices into 10 equally-sized communities, we vary the ground-truth latitude $\ell(\boldsymbol{b}(T))$ between 0 and π and take the input vector $\boldsymbol{x} = \boldsymbol{r}(T)$ with the standard deviation chosen such that $d_{\mathrm{cc}}(\boldsymbol{r}(T), \boldsymbol{b}(T)) \approx \pi/3$. We then apply our heuristic and compare the resulting candidate latitude to the solutions λ_- and λ_+. We see that the latitude of the candidate clustering is indeed much closer to λ_- than λ_+, except when $\lambda_- \approx \lambda_+$, which occurs around $\ell(\boldsymbol{b}(T)) = \pi/2$. We formulate the observed behavior in the next empirical observation:

Observation 2. *The latitude of the candidate clustering that results from our heuristic is closer to λ_- than to λ_+, except possibly when $\lambda_- \approx \lambda_+$.*

We will now prove our main result:

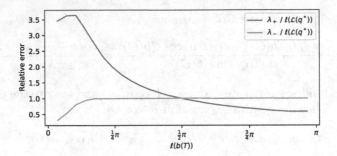

Fig. 3. Comparing the two possible candidate latitudes from Lemma 2 to the empirical candidate latitude that results from our heuristic. We take $\boldsymbol{x} = \boldsymbol{r}(T)$ and choose the noise vector such that $d_{\mathrm{CC}}(\boldsymbol{r}(T), \boldsymbol{b}(T)) \approx \pi/3$, and vary $\ell(\boldsymbol{b}(T))$.

Theorem 1. *Let $\boldsymbol{q}^*(\boldsymbol{x})$ be the query vector that results from projecting the input vector $\boldsymbol{x}$ to the heuristic latitude given by (8). If Observation 2 holds, and Observation 1 holds with equality, then the candidate clustering and the true clustering have the same latitude. That is,*

$$d_a(\boldsymbol{q}^*, \mathcal{L}(\boldsymbol{q}^*)) = d_{\mathrm{CC}}(\boldsymbol{q}^*, \mathcal{L}(\boldsymbol{q}^*)) = \theta \quad implies \quad \ell(\mathcal{L}(\boldsymbol{q}^*)) = \ell(\boldsymbol{b}(T)).$$

Proof. Lemma 1 tells us that for $\boldsymbol{q}^*$, $d_a(\boldsymbol{q}^*, \boldsymbol{b}(T)) = d_{\mathrm{CC}}(\boldsymbol{q}^*, \boldsymbol{b}(T))$ holds. The last condition to check before Observation 1 applies is $\lambda^* > \theta$, or equivalently, $\cos \lambda^* < \cos \theta$. Since $\cos \theta > 0$ and $\sin \lambda_T \sin \theta > 0$, we have

$$\cos \lambda^* = \frac{\cos \lambda_T \cos \theta}{1 + \sin \lambda_T \sin \theta} < \cos \theta.$$

Thus, $\boldsymbol{q}^*$ indeed satisfies the conditions of Observation 1. As stated in the theorem, we assume that Observation 1 holds with equality:

$$d_a(\boldsymbol{q}^*, \mathcal{L}(\boldsymbol{q}^*)) = d_{\mathrm{CC}}(\boldsymbol{q}^*, \mathcal{L}(\boldsymbol{q}^*)) = \theta.$$

Thus, Lemma 2 applies. Together with Observation 2, this gives

$$\cos \ell(\mathcal{L}(\boldsymbol{q}^*)) = \frac{\cos \lambda^* \cos \theta}{1 - \sin \lambda^* \sin \theta}.$$

Substituting $\cos \lambda^*$ and $\sin \lambda^*$ from (8) and (10), we get

$$\cos \ell(\mathcal{L}(\boldsymbol{q}^*)) = \frac{\cos \lambda_T \cos^2 \theta}{1 + \sin \lambda_T \sin \theta - \sin \lambda_T \sin \theta - \sin^2 \theta} = \cos \lambda_T.$$

Since the cosine is injective on $[0, \pi]$ and $\ell(\mathcal{L}(\boldsymbol{q}^*)), \lambda_T \in [0, \pi]$ by the definition of latitude, this implies that $\ell(\mathcal{L}(\boldsymbol{q}^*)) = \lambda_T$. $\qquad\square$

The result of Theorem 1 is demonstrated in Figs. 1d and 1c. We further demonstrate Theorem 1 using the same experiment as in Fig. 2. The results (see Fig. 4) show that $\ell(\boldsymbol{b}(C)) \approx \ell(\boldsymbol{b}(T))$ indeed holds for $\boldsymbol{b}(C) = \mathcal{L}(\boldsymbol{q}^*)$.

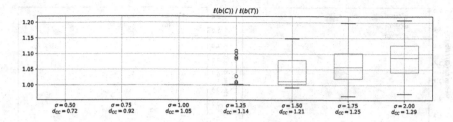

Fig. 4. Validating Theorem 1 for the query vector $\boldsymbol{x} = \boldsymbol{r}(T)$ for different noise levels. The "$d_{CC} = \dots$" under the plots refer to the correlation distances between $\boldsymbol{r}(T)$ and $\boldsymbol{b}(T)$ corresponding to the noise levels.

Unknown Parameters and Estimate for λ^*. To compute the heuristic latitude of (8), one needs some knowledge of $\boldsymbol{b}(T)$. Namely, one needs to know the latitude and the correlation distance to the input vector. As derived in Lemma 1, this is the query latitude for which $d_a(\boldsymbol{q}^*, \boldsymbol{b}(T)) = d_{CC}(\boldsymbol{q}^*, \boldsymbol{b}(T))$ holds. However, Observation 2 tells us that this is approximately equal to the intersection of the curves $\lambda \mapsto d_a(\boldsymbol{q}(\lambda), \mathcal{L}(\boldsymbol{q}(\lambda)))$ and $\lambda \mapsto d_{CC}(\boldsymbol{q}(\lambda), \mathcal{L}(\boldsymbol{q}(\lambda)))$ for $\boldsymbol{q}(\lambda) = \mathcal{P}_\lambda(\boldsymbol{x})$. Therefore, one can consider the intersection $\hat{\lambda}$ of these curves as an estimate of λ^*, and computing this estimate does not require any knowledge of $\boldsymbol{b}(T)$. The evaluation of this estimate is beyond the scope of this paper.

5   Experiments

We will now experimentally demonstrate that our heuristic produces a candidate clustering of a correct latitude, and performs on par with, or better than, the MLE and modularity maximization methods.

All experiments are performed on synthetic networks so that we know the ground truth, and the MLE methods are based on the correct model, so that they give a justifiable baseline.

Our first experiment compares the latitude and performance of our heuristic on PPM to MLE and the standard ER-modularity as a function of the fraction of inter-community edges. We consider graphs of 1000 vertices with 50 equally-sized communities and average degree 8, while we let this fraction vary between 0.05 and 0.65. For each value of this fraction, we generate 50 graphs. In Fig. 5a, we plot the median ratio between the latitude of the candidate and the ground-truth clustering vectors. A ratio close to 1 means that the candidate clustering has a latitude close to the correct latitude. We see that our heuristic with input vector $e(G)$ preserves the latitude better than MLE and ER-modularity. Interestingly, MLE tends to result in clusterings of a lower latitude (thus, smaller communities), while the ER-modularity returns a candidate clustering of a higher latitude (thus, larger communities). The latter tendency of modularity maximization is well-known in the literature [3,23]. In Fig. 5b we see that, despite the fact that MLE uses the correct model, our heuristic performs comparable and sometimes even outperforms MLE.

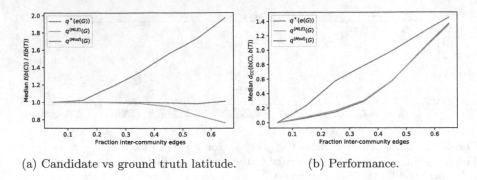

(a) Candidate vs ground truth latitude.   (b) Performance.

Fig. 5. MLE, ER-modularity, and our heuristic with input vector $e(G)$, applied to PPM graphs.

In our final experiment, we compare the performance of our heuristic for several input vectors to other methods, on three random graphs, 1000 vertices each. In each graph, the average degree is approximately 8 and roughly a quarter of the edges are inter-community edges. The Planted Partition Model (PPM) and the Degree Corrected Planted Partition Model (DCPPM) consist of 50 communities of size 20 each. In the heterogeneously-sized PPM (HPPM), the community sizes follow a power-law distribution with exponent 2.5, that is, the probability that a community has size s is proportional to $s^{-2.5}$. In DCPPM, the degrees follow a power-law distribution with exponent 2.5, while the degrees in PPM and HPPM are (approximately) Poisson distributed. For the Artificial Benchmark for Community Detection (ABCD, [10]), we take a degree sequence with power-law exponent 2.5 and community sizes with power-law exponent 1.5. We set the remaining parameters of ABCD so that the average degree and the fraction of inter-community edges are similar to the other models[3]. We generate each random graph 50 times and present the median of the results in Table 1.

Table 1. Median of $d_{CC}(b(C), b(T))$ over 50 samples of each model for several projection methods. The best score is written in bold for each graph model.

	$q^{(\mathrm{MLE})}(G)$	$q_{\mathrm{DC}}^{(\mathrm{MLE})}(G)$	$q^*(e(G))$	$q^*(q_{\mathrm{DC}}^{(\mathrm{MLE})}(G))$	$q^*(w(G))$	$q^*(j(G))$
PPM	0.1633	0.1583	**0.1440**	0.1514	0.1793	0.1444
HPPM	0.6537	0.6236	0.5758	0.5500	0.5161	**0.4625**
DCPPM	0.5512	**0.5274**	0.5419	0.5434	1.1875	0.9048
ABCD	0.6018	0.3278	0.5821	**0.1933**	0.8526	0.5830

---

[3] We use $\xi = 0.25$, degree bounds 4 and 100, community-size bounds 10 and $\lfloor n^{3/4} \rfloor$.

For PPM graphs, we surprisingly see that the MLE of DCPPM performs better (0.1583) than the true MLE (0.1633). Applying our heuristic using PPM-MLE as input vector improves the performance even more (0.1440), while the Jaccard vector (0.1444) achieves similar performance. For the heterogeneously-sized PPMs, we see that all four variants of our heuristic outperform the MLE methods, with the Jaccard input vector (0.4625) achieving the best performance. For the DCPPM graphs, we see that the MLE (0.5274) performs best and this is the only instance in Table 1 where applying our heuristic to an MLE input vector does not lead to improved performance. Finally, for the ABCD graphs, we see that our heuristic applied to the DCPPM-MLE input vector (0.1933) significantly outperforms the other methods.

Overall, we see that in almost all cases (except for DCPPM-MLE on DCPPM graphs), applying our heuristic to the MLE query vector improves the performance. Note that the reported values are the medians over 50 trials. For individual graphs, it may still occur that the MLE performs better than our heuristic. However, as shown in the previous experiment, our heuristic has the advantage of a much more accurate value for the latitude of the resulting clustering vector.

Finally, we perform an experiment on larger ABCD graphs to demonstrate the scalability of projection methods in Tables 2 and 3. Comparing the first two columns of Table 2, we see that applying our heuristic to the DCPPM-MLE query vector greatly improves the performance, while the Jaccard and wedge vectors perform worse. In Table 3, we see that the running times for the query vectors $q_{DC}^{(MLE)}(G)$ and $q^*(q_{DC}^{(MLE)}(G))$ are comparable, which tells us that the heuristic does not significantly affect the running time. We also see that the query vectors $q^*(w(G))$ and $q^*(j(G))$ have a significantly higher running time. This shows that the choice of the query vector greatly affects the running time of the projection method, as discussed in [4].

Table 2. Performance of several projection methods on ABCD graphs of different sizes. For each n, we sample one ABCD graph, apply the projection methods and report the $d_{CC}(b(C), b(T))$ obtained by each projection method. The results of $q^*(w(G))$ and $q^*(j(G))$ for $n = 100,000$ are omitted as they took longer than a day to compute.

n	$q_{DC}^{(MLE)}(G)$	$q^*(q_{DC}^{(MLE)}(G))$	$q^*(w(G))$	$q^*(j(G))$
1,000	0.45	**0.29**	0.97	0.53
10,000	0.42	**0.16**	1.02	0.85
100,000	0.29	**0.12**	-	-

Table 3. Running times (in seconds) of the projection methods on ABCD graphs from Table 2.

n	$q_{\mathrm{DC}}^{(\mathrm{MLE})}(G)$	$q^*(q_{\mathrm{DC}}^{(\mathrm{MLE})}(G))$	$q^*(w(G))$	$q^*(j(G))$
1,000	3.0	3.1	27.2	18.9
10,000	51.7	60.9	1,473.9	1,262.9
100,000	2,590.9	2,993.5	-	-

6   Discussion

We have introduced a heuristic that allows to correct for the granularity bias from which modularity-based methods typically suffer. This heuristic can be applied to any type of pair-wise similarity data. When using edge-connectivity as similarity, we recover the ER-modularity maximization. In Sect. 5, we saw that applying our heuristic to other similarity data as input vector may result in community detection methods that clearly outperform likelihood-based methods while preserving the clustering granularity close to the ground truth.

Our experiments show that there is not a single best-performing input vector, but that it depends on the graph at hand. In [4], we suggested that low values of the correlation distance between the input vector and the ground truth typically lead to good performance. However, by this criterion, most of the time, input vector $w(G)$ should outperform $e(G)$, while we did not observe this in the experiments. Thus, the choice of a suitable input vector largely remains an open question.

While we were able to derive our heuristic from empirical observations, we were unable to theoretically explain the observed phenomena. This is mainly due to the lack of analytical methods for predicting outcomes of the Louvain algorithm. A way to overcome this difficulty might be to replace the Louvain algorithm by a different algorithm that is easier to analyze (see, e.g., [14]). While there exist theoretical results regarding the modularity maxima of several random graph models [9,13,14,20], these results apply to CM-modularity with resolution 1 and extending these to other resolution parameter values is nontrivial.

We emphasize that our heuristic has a low computational complexity. Indeed, the complexity of our heuristic is similar to the one of modularity maximization. In this work, the heaviest computational tasks were to compute the Louvain projection for the wedge and Jaccard vectors.

Finally, in further research we want to apply our methods to large real-life networks. As we have discussed in Sect. 4, not knowing the ground truth can be overcome in practice, although the method we have proposed for this requires further evaluation. For the sake of this paper, we stick to synthetic networks mainly due to the lack of data with known ground truths and many communities. The former is needed because the goal of this paper is to validate the heuristic. The latter is important because our heuristic is especially advantageous in the realistic situation when the granularity of the ground truth is not very coarse.

References

1. Arenas, A., Fernandez, A., Gomez, S.: Analysis of the structure of complex networks at different resolution levels. New J. Phys. **10**(5), 053039 (2008)
2. Blondel, V.D., Guillaume, J.L., Lambiotte, R., Lefebvre, E.: Fast unfolding of communities in large networks. J. Stat. Mech: Theory Exp. **2008**(10), P10008 (2008)
3. Fortunato, S., Barthélemy, M.: Resolution limit in community detection. Proc. Natl. Acad. Sci. **104**(1), 36–41 (2007). https://doi.org/10.1073/pnas.0605965104. https://www.pnas.org/content/104/1/36
4. Gösgens, M., van der Hofstad, R., Litvak, N.: The hyperspherical geometry of community detection: modularity as a distance. J. Mach. Learn. Res. **24**, 1–33 (2023)
5. Gösgens, M., Zhiyanov, A., Tikhonov, A., Prokhorenkova, L.: Good classification measures and how to find them. Adv. Neural. Inf. Process. Syst. **34**, 17136–17147 (2021)
6. Gösgens, M.M., Tikhonov, A., Prokhorenkova, L.: Systematic analysis of cluster similarity indices: how to validate validation measures. In: International Conference on Machine Learning, pp. 3799–3808. PMLR (2021)
7. Hunter, P.R., Gaston, M.A.: Numerical index of the discriminatory ability of typing systems: an application of simpson's index of diversity. J. Clin. Microbiol. **26**(11), 2465–2466 (1988)
8. Jaccard, P.: The distribution of the flora in the alpine zone. 1. New Phytol. **11**(2), 37–50 (1912)
9. Kamiński, B., Pankratz, B., Prałat, P., Théberge, F.: Modularity of the ABCD random graph model with community structure. J. Complex Netw. **10**(6), cnac050 (2022)
10. Kamiński, B., Prałat, P., Théberge, F.: Artificial benchmark for community detection (ABCD)-fast random graph model with community structure. Netw. Sci. **9**(2), 153–178 (2021)
11. Kumpula, J.M., Saramäki, J., Kaski, K., Kertész, J.: Limited resolution in complex network community detection with Potts model approach. Eur. Phys. J. B **56**(1), 41–45 (2007)
12. Lancichinetti, A., Fortunato, S.: Limits of modularity maximization in community detection. Phys. Rev. E **84**(6), 066122 (2011)
13. Lichev, L., Mitsche, D.: On the modularity of 3-regular random graphs and random graphs with given degree sequences. Random Struct. Algorithms **61**(4), 754–802 (2022)
14. McDiarmid, C., Skerman, F.: Modularity of erdős-rényi random graphs. Random Struct. Algorithms **57**(1), 211–243 (2020)
15. Newman, M.E.: Equivalence between modularity optimization and maximum likelihood methods for community detection. Phys. Rev. E **94**(5), 052315 (2016)
16. Newman, M.E., Girvan, M.: Finding and evaluating community structure in networks. Phys. Rev. E **69**(2), 026113 (2004)
17. Peixoto, T.P.: Descriptive vs. inferential community detection: pitfalls, myths and half-truths. arXiv preprint arXiv:2112.00183 (2021)
18. Prokhorenkova, L.: Using synthetic networks for parameter tuning in community detection. In: Avrachenkov, K., Prałat, P., Ye, N. (eds.) WAW 2019. LNCS, vol. 11631, pp. 1–15. Springer, Cham (2019). https://doi.org/10.1007/978-3-030-25070-6_1

19. Prokhorenkova, L., Tikhonov, A.: Community detection through likelihood optimization: in search of a sound model. In: The World Wide Web Conference, pp. 1498–1508 (2019)
20. Ostroumova Prokhorenkova, L., Prałat, P., Raigorodskii, A.: Modularity of complex networks models. In: Bonato, A., Graham, F.C., Prałat, P. (eds.) WAW 2016. LNCS, vol. 10088, pp. 115–126. Springer, Cham (2016). https://doi.org/10.1007/978-3-319-49787-7_10
21. Rand, W.M.: Objective criteria for the evaluation of clustering methods. J. Am. Stat. Assoc. **66**(336), 846–850 (1971)
22. Reichardt, J., Bornholdt, S.: Statistical mechanics of community detection. Phys. Rev. E **74**(1), 016110 (2006)
23. Traag, V.A., Van Dooren, P., Nesterov, Y.: Narrow scope for resolution-limit-free community detection. Phys. Rev. E **84**(1), 016114 (2011)
24. Zhang, L., Peixoto, T.P.: Statistical inference of assortative community structures. Phys. Rev. Res. **2**(4), 043271 (2020)

The Emergence of a Giant Component in One-Dimensional Inhomogeneous Networks with Long-Range Effects

Peter Gracar[1] , Lukas Lüchtrath[2]([envelope]) , and Christian Mönch[3]

[1] University of Leeds, Leeds, UK
P.Gracar@leeds.ac.uk
[2] Weierstrass Institute for Applied Analysis and Stochastics, Berlin, Germany
lukas.luechtrath@wias-berlin.de
[3] Johannes Gutenberg Universität Mainz, Mainz, Germany
cmoench@uni-mainz.de

Abstract. We study the weight-dependent random connection model, a class of sparse graphs featuring many real-world properties such as heavy-tailed degree distributions and clustering. We introduce a coefficient, δ_{eff}, measuring the effect of the degree-distribution on the occurrence of long edges. We identify a sharp phase transition in δ_{eff} for the existence of a giant component in dimension $d = 1$.

Keywords: Long-range effects · Percolation · Phase transition · Spatial random graphs · Preferential attachment · Boolean model

1 Introduction and Statement of Result

Complex real-world systems can be seen as a collection of numerous objects interacting with each other in specific ways. This holds in many different contexts and fields such as biology, physics, telecommunications, social sciences, information technology and more. Put differently, many complex systems can be seen as a network where the objects are described by the network's nodes and a link between two nodes in the network indicates the interaction between the corresponding objects. Therefore, over the last 20 years complex networks have become a key tool used to describe real-world systems and related problems. Despite the large amount of uncertainty and complexity arising from their dynamical nature, it is of great importance to understand the structure of the underlying network when analysing a such a system. What kind of phenomena arise in the system and how can they be explained by the way the network is built? These are typical questions in the scientific community but which are also of public interest as their answers may affect decisions made by political or economic leaders.

In recent years, the increase in computing power has made more and more real-world networks amenable to empirical analysis. Most interestingly, despite their different contexts, many such networks have similar structural properties,

M. Dewar et al. (Eds.): WAW 2023, LNCS 13894, pp. 19–35, 2023.
https://doi.org/10.1007/978-3-031-32296-9_2

see e.g. [5,8]. Can these common structural features be explained in a simple way by basic local principles which are shared by the different networks? Such mechanisms are for instance [13]:

- Networks contain nodes that are far more influential than an average node; the so called *hubs* or *stars*.
- Networks show strong *clustering*: Nodes sharing a common neighbour are much more likely to be connected by an edge themselves than nodes that are picked randomly.

Put differently, nodes prefer to connect to similar nodes or to highly influential ones. Despite this, links between nodes that are neither similar nor influential should still occasionally arise. We are interested in models based on these building principles and how classical properties of networks such as the degree distribution, the size of the largest connected component or typical graph distances are affected by them.

In this article, we present a class of models where the vertices are embedded into Euclidean space and each vertex is given an independent weight. Here, the spatial location of a vertex abstractly represents some intrinsic feature and spatially close vertices can be seen as similar and eager to connect to each other. The weight of a vertex represents its influence within the system. The connection mechanism then depends on both weight and spatial distance, hence connections to spatially close vertices or vertices with a large weight are more probable. The introduction of weights guarantees heavy-tailed degree distributions leading to the existence of hubs. The spatial embedding leads to clustering. However, the connection mechanism is set up in a way that still allows far apart vertices having only typical weights to occasionally connect. We call connections of the latter type *long-range* connections. We introduce a coefficient, δ_{eff}, depending only on basic model parameters, which quantifies the overall occurrence of long-range connections in a way that makes it comparable to classical long-range percolation without weights. In this standard model edges are present independently from each other with a probability decaying polynomially in the distance of the endpoints and it corresponds to a homogeneous version of our model. In the present work we focus on the question of existence of a giant component in one dimension, that is, a connected component whose size is of the same order as the entire network. The geometric restrictions of one dimensional space make the existence of a giant rather difficult to achieve. We shall see that the behaviour of long-range percolation is paradigmatic for the models involving weights when expressed in terms of δ_{eff}. Although we focus primarily on one dimension, δ_{eff} also plays a significant role in higher dimensions, for example in providing a sufficient criterion for the existence of subcritical percolation phases [24]. We further believe that it also provides a sufficient criterion for transience of the infinite limit of the giant cluster.

We note that δ_{eff} is a purely theoretical value determined by the network topology that characterises the behaviour of certain useful properties as outlined above. At this stage we are unable to provide any approaches for estimating δ_{eff} using observed data. Any such method would need to take into account that δ_{eff}

inherently involves multiple scales, as it describes the decay of edge probabilities between aggregate vertex sets on increasingly larger scales. This contrasts with the homogeneous case, in which the decay exponent can be estimated directly from the empirical edge length distribution. We believe that estimating δ_{eff} is an interesting statistical problem with potential applications in the analysis of large scale networks.

1.1   The Weight-Dependent Random Connection Model

The *weight-dependent random connection model* is a class of infinite graphs on the points of a Poisson point process on $\mathbb{R}^d \times (0,1)$ that has been intensively studied in recent years [15–17]. In the present article we introduce a finite version of the weight-dependent random connection model constructed on the unit torus $\mathbb{T}_1^d = (-1/2, 1/2)^d$ equipped with torus metric

$$d_1(x,y) = \min\{|x - y + u| : u \in \{-1,0,1\}^d\}, \text{ for } x,y \in \mathbb{T}_1^d$$

to avoid boundary effects. Here and throughout the paper $|\cdot|$ denotes the Euclidean norm. For $N > 0$ and $\beta > 0$, we construct the graph $\mathscr{G}_N^\beta$ in the following way: The vertex set of $\mathscr{G}_N^\beta$ is a Poisson point process $\mathcal{X}_N$ of intensity N on $\mathbb{T}_1^d \times (0,1)$. We denote a vertex by $\mathbf{x} = (x, t_x) \in \mathcal{X}_N$ and call $x \in \mathbb{T}_1^d$ the *location* and $t_x \in (0,1)$ the *mark* of the vertex. Given $\mathcal{X}_N$, each pair of vertices $\mathbf{x} = (x, t_x)$ and $\mathbf{y} = (y, t_y)$ is connected independently by an edge with probability

$$1 \wedge \left(\tfrac{1}{\beta}(t_x \wedge t_y)^\gamma (t_x \vee t_y)^\alpha N\, d_1(x,y)^d \right)^{-\delta}, \tag{1}$$

for $\gamma \in [0,1), \alpha \in [0, 2 - \gamma)$ and $\delta > 1$ where we denote with $a \wedge b$ the minimum and with $a \vee b$ the maximum of a and b.

Remark 1.

(i) As the typical distance of a point to its nearest neighbour in $\mathcal{X}_N$ is of order $N^{-1/d}$, it is necessary to scale the distance by it to avoid that the graph degenerates. Note that in law it is the same to construct the graph on the points of a unit intensity Poisson process on the volume N torus $\mathbb{T}_N^d = (-N^{1/d}/2, N^{1/d}/2)^d$ and replace $N\, d_1(x,y)^d$ in (1) by $d_N(x,y)^d$, the torus metric of $\mathbb{T}_N^d$.

(ii) The parameter $\beta > 0$ controls the edge intensity. Since $x \mapsto 1 \wedge x^{-\delta}$ is a non increasing function, a larger value of β increases the connection probability which then leads to more edges on average.

(iii) The parameters γ and α control the way of influence that vertex marks have on the connection mechanism. By construction, connections to vertices with a small mark are preferred, see also Fig. 2 below. The mark can therefore be seen as the inverse weight of a vertex, giving the model its name. Various choices of γ and α lead to (finite versions of) various models established in the literature, see Table 1.

Table 1. Various choices for γ, α and δ and the models they represent in their infinite version. Here, to shorten notation, $\delta = \infty$ represents models constructed with a function ρ of bounded support, cf. Remark 1(iv).

Parameters	Names and references
$\gamma = 0, \alpha = 0, \delta = \infty$	random geometric graph, Gilbert's disc model [11]
$\gamma = 0, \alpha = 0, \delta < \infty$	random connection model [29,31], long-range percolation [32]
$\gamma > 0, \alpha = 0, \delta = \infty$	Boolean model [12,18], scale-free Gilbert graph [20]
$\gamma > 0, \alpha = 0, \delta < \infty$	soft Boolean model [14]
$\gamma = 0, \alpha > 1, \delta = \infty$	ultra-small scale-free geometric network [33]
$\gamma > 0, \alpha = \gamma, \delta \leq \infty$	scale-free percolation [6,7], geometric inhomogeneous random graphs [4]
$\gamma > 0, \alpha = 1 - \gamma, \delta \leq \infty$	age-dependent random connection model [13]

(iv) The parameter δ controls the occurrence of long edges. The larger the value of δ, the stronger the effect of the geometric embedding is and the less long edges are present. One can replace the function $1 \wedge x^{-\delta}$ in (1) by a non increasing function $\rho : (0, \infty) \to [0, 1]$ and the geometric restrictions become hardest when ρ is of bounded support. Results about such a model can be derived from our model as a limit $\delta \to \infty$.

(v) The restrictions for γ, α and δ guarantee that

$$\int_0^1 ds \int_t^1 dt \int_0^\infty dx \left(1 \wedge (\beta^{-1} s^\gamma t^\alpha x)^{-\delta} \right) < \infty,$$

and therefore all expected degrees remain finite when $N \to \infty$. Consequently, the graph is sparse in the sense that the number of edges is of the same order as the size of the graph.

For all choices of γ, δ and α, the above model converges to a local limit as $N \to \infty$ [13,22], where the limiting graph $\mathscr{G}^\beta = \mathscr{G}^\beta_\infty$ is constructed as follows: The vertex set is given by a unit intensity Poisson process on $\mathbb{R}^d \times (0,1)$ and two given points $\mathbf{x} = (x, t_x)$ and $\mathbf{y} = (y, t_y)$ are connected independently by an edge with probability

$$1 \wedge \left(\tfrac{1}{\beta}(t_x \wedge t_y)^\gamma (t_x \vee t_y)^\alpha |x - y|^d \right)^{-\delta}. \tag{2}$$

Here, the term local limit is to be understood in the following way: Add a vertex $\mathbf{0} = (0, U)$ at the origin to the graph having a uniform mark U and connect it to all other vertices by rule (2). By Palm theory [27, Chapter 9] this is the same as shifting the graph such that a typical vertex is located at the origin. Then for each event $A(\mathbf{0}, \mathscr{G}^\beta_N)$ depending on the origin and a bounded graph neighbourhood of it in $\mathscr{G}^\beta_N$, we have

$$\lim_{N \to \infty} \mathbb{P}(A(\mathbf{0}, \mathscr{G}^\beta_N)) = \mathbb{P}(A(\mathbf{0}, \mathscr{G}^\beta_\infty)).$$

Put differently, when the number of vertices N tends to infinity, the local neighbourhoods in $\mathscr{G}^\beta_N$ and $\mathscr{G}^\beta_\infty$ look the same.

A similar modelling approach only using a different parametrisation is that of "geometric inhomogeneous random graphs" [4,25] and their infinite volume local limits. All appearing parameters in both approaches can be translated from one model into the other [23].

The limiting graph can then directly be used to derive results for the (asymptotic) degree distribution and clustering since both depend only on graph neighbourhoods of length at most two. The following theorem summarises results from [13,28].

Theorem 1 (Degree distribution and clustering). *Let $\mathscr{G}_N^\beta$ be the weight-dependent random connection model for some choice of $\delta > 1$, $\gamma \in [0,1)$ and $\alpha \in [0, 2 - \gamma)$.*

(i) There exists a probability sequence $(\mu_k : k \geq 0)$ such that in probability

$$\lim_{N \to \infty} \frac{1}{N} \sum_{\mathbf{x} \in \mathscr{G}_N^\beta} \mathbb{1}_{\{\mathbf{x} \text{ has } k \text{ neighbours in } \mathscr{G}_N^\beta\}} = \mu_k.$$

Moreover, for $\tau := 1 + (1/\gamma \wedge 1/(\gamma + \alpha - 1)^+)$, we have

$$\lim_{k \to \infty} k^{\tau + o(1)} \mu_k = 1.$$

(ii) Denote by $V_2(\mathscr{G}_N^\beta)$ the set of vertices with at least two neighbours in $\mathscr{G}_N^\beta$. If $\mathbf{y}$ and $\mathbf{z}$ are neighbours of a vertex $\mathbf{x}$, we call $\{\mathbf{x}, \mathbf{y}, \mathbf{z}\}$ a triangle when also $\mathbf{y}$ and $\mathbf{z}$ are connected by an edge. Then, there exists a positive constant c depending only on the model parameters such that in probability

$$\lim_{N \to \infty} \frac{1}{N} \sum_{\mathbf{x} \in V_2(\mathscr{G}_N^\beta)} \frac{\sharp\{\text{triangles containing } \mathbf{x}\}}{\binom{\sharp\{\text{neighbours of } \mathbf{x}\}}{2}} = c.$$

Theorem 1(i) shows that the degree distribution is heavy-tailed and therefore $\mathscr{G}_N^\beta$ contains the aforementioned hubs. Part (ii) shows that $\mathscr{G}_N^\beta$ indeed exhibits clustering.

From a modelling point of view, the weight-dependent random connection model has the huge advantage that it allows a large flexibility in modelling the way the weight influence the networks geometry. Moreover, sparseness of the graph together with the conditional independence and the ranking of the vertices by their marks can be used to construct the graph in linear time. The following result is an adaption of [3].

Theorem 2. *If $\gamma > 0$, then $\mathscr{G}_N^\beta$ can be sampled in time $O(N)$.*

1.2   Main Result

More difficult than deriving the degree distribution or clustering of a network is the question of the existence of a connected component of linear size. More precisely, we say that $\mathscr{G}_N^\beta$ contains a *giant component* if

$$\lim_{\varepsilon \downarrow 0} \limsup_{N \to \infty} \mathbb{P}(\sharp \mathscr{C}(\mathscr{G}_N^\beta) < \varepsilon N) = 0,$$

where $\mathscr{C}(\mathscr{G}_N^\beta)$ denotes the largest connected component of $\mathscr{G}_N^\beta$ and $\sharp\mathscr{C}(\mathscr{G}_N^\beta)$ the number of vertices within. Note that the existence of a giant component does not only depend on bounded graph neighbourhoods but on the entire graph. Therefore, the local limit structure cannot be used directly. However, it is known that the existence of a giant is highly linked with the existence of an infinite component in the limiting graph. Only when the limiting graph contains an infinite component, a giant component can exist [21]. Our main theorem concerns the existence of a giant component in dimension $d = 1$ where the existence of infinite components in the limit is particularly hard due to the restrictions of the real line $\mathbb{R}$. For the standard long-range percolation model, i.e. the local limit of our model for the choice of $\alpha = \gamma = 0$, it is known that this question depends on the occurrence of long-edges, measured by δ. More precisely, there exists an infinite connected component in the limit for large enough β if $\delta \leq 2$ but there does not exist such a component for all β if $\delta > 2$ [1,30]. We now introduce the *effective decay exponent* δ_{eff}, measuring the influence of the vertex weights on the occurrence of long edges, as

$$\delta_{\text{eff}} := \lim_{n\to\infty} \frac{\log \int_{1/n}^1 ds \int_s^1 dt \left(1 \wedge (s^\gamma t^\alpha n)^{-\delta}\right)}{\log n}. \tag{3}$$

Theorem 3 (Existence vs. non existence of a giant). *Let $\mathscr{G}_N^\beta$ be the weight-dependent random connection model in dimension $d = 1$.*

(i) *If $\delta_{\text{eff}} < 2$, then the network $\mathscr{G}_N^\beta$ contains a giant component for large enough values of β.*

(ii) *If $\delta_{\text{eff}} > 2$, then the limiting graph $\mathscr{G}_\infty^\beta$ does not contain an infinite component for any value of β and no giant component can exist in $\mathscr{G}_N^\beta$.*

1.3   Examples

In this section we present and further discuss two particularly interesting examples covered by our framework.

Age-Based Spatial Preferential Attachment. This model can be seen as the most natural model in our framework. It corresponds to the choice of $\gamma > 0$ and $\alpha = 1 - \gamma$. Its local limit is known under the name *age-dependent random connection model* [13] and it is a type of preferential attachment model. In it, the marks represent the vertices' birth times and early birth times correspond to old hence present for a long time vertices. As $\gamma + \alpha = 1$, one can rewrite the connection probability (1) by

$$1 \wedge \left(\tfrac{1}{\beta}(Nt_x \wedge Nt_y)^\gamma (Nt_x \vee Nt_y)^{1-\gamma} \, d_1(x,y)^d\right)^{-\delta}$$

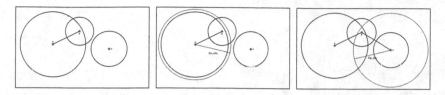

Fig. 1. Example for the connection mechanism of the soft Boolean model in two dimensions. The solid lines represent the graph's edges.

and hence the model can also be constructed on a unit intensity Poisson process on $\mathbb{T}_1^d \times (0, N)$. In this situation, vertices arrive after standard exponential waiting times, justifying the notion of marks being birth times. Since

$$\delta_{\text{eff}} \begin{cases} < 2, & \gamma > 1 - \frac{1}{\delta}, \\ = 2, & \gamma \leq 1 - \frac{1}{\delta} \end{cases},$$

the graph contains a giant component for sufficiently large β when $\gamma > 1 - 1/\delta$. From [16], we can derive that the critical value β_c after which a giant exists is larger than zero for $\gamma < 1 - 1/(\delta+1)$ and zero for $\gamma > 1 - 1/(\delta+1)$. The case $\delta_{\text{eff}} = 2$ is the boundary case which is not covered by our main theorem. However, for $\gamma = 1/2$ the values of γ and α coincide and the minimum and maximum structure in (2) reduces to the product $\sqrt{t_x t_y}$. Therefore, our model coincides with a hyperbolic random graph model [26] for which the existence of a giant is known for large enough β [2]. Hence, by a domination argument, the considered model contains a giant component for all $\gamma \geq 1/2$. It remains an interesting open problem whether this remains true for $\gamma \in (0, 1/2)$.

Soft Boolean Model. This model corresponds to the choice of $\gamma > 0$ and $\alpha = 0$. Following the representation of [14], each vertex x is assigned an independent radius $R_x := t_x^{-\gamma/d}$. Additionally, each potential edge $\{x, y\}$ is assigned an independent random variable $Z(x, y)$ with tail-distribution function $\mathbb{P}(Z(x, y) \geq z) = 1 \wedge z^{-\delta}$. Given the vertices and the collection of $Z(x, y)$, two vertices in $\mathscr{G}_N^\beta$ are connected by an edge, when

$$N^{1/d} \, \mathrm{d}_1(x, y) \leq \beta^{1/d} Z(x, y) (R_x \vee R_y).$$

That is, the vertices share an edge when in the rescaled picture (cf. Remark 1(i)), the vertex with smaller assigned radius is contained in the ball centered at the stronger vertex with the assigned radius stretched by a heavy-tailed random variable, cf. Fig. 1. For $\delta \to \infty$ one derives a version of the classical Boolean model [20]. We calculate

$$\delta_{\text{eff}} = 1 + \delta(1 - \gamma) \begin{cases} < 2, & \gamma > 1 - \frac{1}{\delta}, \\ = 2, & \gamma = 1 - \frac{1}{\delta}, \\ > 2, & \gamma < 1 - \frac{1}{\delta}. \end{cases}$$

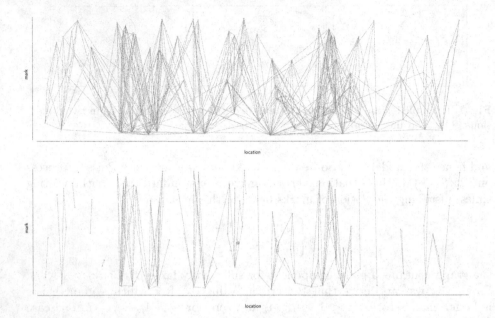

Fig. 2. Two simulations for the (rescaled) soft Boolean model on the torus $\mathbb{T}_{100}^1$ with $\delta = 3.5$ and $\beta = 1$. The first picture was simulated with $\gamma = 0.75$, the second one with $\gamma = 0.51$.

Therefore, the graph $\mathscr{G}_N^\beta$ contains a giant component for large enough values of β for $\gamma > 1 - 1/\delta$ but does not for $\gamma < 1 - 1/\delta$, see also Fig. 2 for some simulations.

2   Proof of the Main Theorem

2.1   Some Construction and Notation

From now on, we work exclusively in dimension $d = 1$. We also work on the rescaled picture based on Remark 1(i). Then the underlying vertex set of our graphs can be constructed in the following way. We start with a vertex $X_0 = 0$ placed at the origin. Let $(Z_i : i \in \mathbb{N})$ and $(\tilde{Z}_i : i \in \mathbb{N})$ be two independent sequences of independent standard exponential random variables. For $i \in \mathbb{N}$ set $X_i = \sum_{j=1}^i Z_j$ and for $i \in \mathbb{Z}\backslash\mathbb{N}_0$ set $X_i = -\sum_{j=1}^{|i|} \tilde{Z}_j$. Then $\eta_0 := \{X_i : i \in \mathbb{Z}\}$ is the Palm version [27] of a unit intensity Poisson process on the real line where a distinguished vertex is placed at the origin. We call η_0 the *vertex locations*. Note that we have $X_i < X_j$ whenever $i < j$. Further, let $\mathcal{T}_0 = (T_i : i \in \mathbb{Z})$ be a sequence of independent Uniform$(0, 1)$ random variables, independent of η_0. The elements of $\mathcal{T}_0$ are the *vertex marks* and we define

$$\mathcal{X}_0 = \left(\mathbf{X}_i = (X_i, T_i) \in \eta_0 \times \mathcal{T}_0 : i \in \mathbb{Z} \right)$$

the Palm version of the marked Poisson process which is our vertex set. Moreover, define $\mathcal{U}_0 = (U_{i,j} : i < j \in \mathbb{Z})$ another sequence of independent Uniform$(0, 1)$ random variables, which we call *edge marks*. Finally define for all $i < j \in \mathbb{Z}$ the random variables

$$E_{i,j} := E_{i,j}(\mathcal{X}_0, \mathcal{U}_0) := \mathbb{1}_{\{U_{i,j} < (\beta^{-1}(T_i \wedge T_j)^\gamma (T_i \vee T_j)^\alpha)|X_i - X_j|)^{-\delta}\}}.$$

Note that the random variables $(E_{i,j} : i < j \in \mathbb{Z})$ are only conditionally independent given $\mathcal{X}_0$ but not in general. The graph $\mathscr{G}^\beta := \mathscr{G}^\beta_\infty(\mathcal{X}_0, \mathcal{U}_0)$ is then defined through its vertex set $\mathcal{X}_0$ and (random) adjacency matrix $(E_{i,j} : i < j \in \mathbb{Z})$.

As a Poisson process remains a Poisson process when restricted to an area, the finite graphs $\mathscr{G}^\beta_N$ can be constructed by the reduction of $\mathcal{X}_0$ and $\mathcal{U}_0$ to the elements in $(-N/2, N/2)$. Note that we may have changed the distance in the connection probability from the torus metric to the Euclidean one. However, Theorem 3(ii) only concerns the infinite model where no change is made and in Theorem 3(i) edges may be removed from the graph due to the larger distances at the boundary which makes the existence of a giant less likely.

Throughout the remaining text, we use the notation $\mathbf{X}_i \sim \mathbf{X}_j$ to denote the event that $\mathbf{X}_i$ and $\mathbf{X}_j$ are connected by an edge. For two sets of vertices V_1, V_2, we write $V_1 \sim V_2$, if they are connected by a direct edge, i.e. there exist $\mathbf{X} \in V_1$ and $\mathbf{Y} \in V_2$ such that $\mathbf{X} \sim \mathbf{Y}$. For two positive functions f and g, we write $f \asymp g$ when f/g is bounded away from zero and infinity. We write $f = o(g)$ if $f(x)/g(x) \to 0$ and $f = O(g)$ if $f(x)/g(x) \to C < \infty$ when $x \to \infty$.

Finally, we say that an event A is *increasing*, if the function $\mathbb{1}_A$ increases whenever additional vertices are added to the graph, vertex marks are decreased (and hence the weights are increased) or edge marks are decreased (and hence potentially more edges are added). For two such events A and B, the FKG-inequality [10] in a version of [19] yields

$$\mathbb{P}(A \cap B) = \mathbb{E}[\mathbb{P}(A \cap B \mid \eta)] \geq \mathbb{E}[\mathbb{P}(A \mid \eta)\mathbb{P}(B \mid \eta)] \geq \mathbb{P}(A)\mathbb{P}(B). \tag{4}$$

2.2 Connecting Far Apart Vertex Sets

To understand the role of δ_{eff}, one has to understand the influence of the vertex marks to the occurrence of long edges. These are essential for large components in dimension one as they are needed to overcome 'bad regions' where far less than average vertices and edges are placed. For large n, the minimum of n independent uniform random variables is roughly $1/n$ and consequently the double integral appearing in (3) is essentially the probability of two randomly picked vertices from vertex sets of size n at distance roughly n being connected by an edge. Ignoring additional correlations between edges and treating the random variables $E_{i,j}$ as independent, the expected number of edges is then roughly given by $n^{2-\delta_{\text{eff}}}$ which grows large for $\delta_{\text{eff}} < 2$ but vanishes for $\delta_{\text{eff}} > 2$. This heuristic is justified in the following lemma which is formulated for one dimension but can be generalised to higher dimensions. For its formulation we define the two sets of vertices

$$V_\ell^n := \{\mathbf{X}_{-2n}, \dots, \mathbf{X}_{-n-1}\} \quad \text{and} \quad V_r^n := \{\mathbf{X}_n, \dots, \mathbf{X}_{2n-1}\}$$

which play the role of the two sets of size n at distance roughly n.

Lemma 1.

(i) For each $\varepsilon > 0$, there exists $\mu \in (0, 1/2)$ and a constant $C > 0$ such that

$$\mathbb{P}(V_\ell^n \not\sim V_r^n) \leq \exp\left(-Cn^{2-\delta_{\text{eff}}-\varepsilon}\right) + O\left(n^{1-\mu}e^{-n^\mu}\right).$$

(ii) For each $\varepsilon > 0$, there exists $\mu \in (0, 1/2)$ and constants $C_1, C_2 > 0$ such that

$$\mathbb{P}(V_\ell^n \not\sim V_r^n) \geq C_1(1 - n^{-\mu})\exp\left(-C_2 n^{2-\delta_{\text{eff}}+\varepsilon}\right)$$

Proof. We start with the proof of (i). To control the influence of the vertex marks and the random distances, we rely on some 'regular' behaviour of the underlying point process. Let $\mu \in (0, 1/2)$ and define $N_\ell^n(i) := \sum_{j=-n-1}^{-2n} \mathbb{1}_{\{T_j \leq i/\lfloor n^{1-\mu} \rfloor\}}$. We say that V_ℓ^n has μ-*regular marks* if

$$N_\ell^n(i) \geq \frac{in}{2\lfloor n^{1-\mu} \rfloor}, \quad \text{for all } i = 1, \ldots, \lfloor n^{1-\mu} \rfloor.$$

Using a standard Chernoff bound for independent uniforms together with the union bound, we infer

$$\mathbb{P}(V_\ell^n \text{ is not } \mu\text{-regular}) = O\left(n^{1-\mu}e^{-n^\mu}\right).$$

The same holds verbatim for V_r^n. Moreover, standard large deviation results for the sum of independent exponential random variables yields

$$\mathbb{P}(|X_{2n-1} - X_{-2n}| > 5n) \leq e^{-\text{const } n}.$$

Therefore, writing $\tilde{\mathbb{P}}$ for $\mathbb{P}$ conditioned of μ-regular marks in V_ℓ^n and V_r^n as well as $|X_{2n-1} - X_{-2n}| \leq 5n$, we have

$$\mathbb{P}(V_\ell^n \not\sim V_r^n) \leq \tilde{\mathbb{P}}(V_\ell^n \not\sim V_r^n) + O\left(n^{1-\mu}e^{-n^\mu}\right). \tag{5}$$

To calculate the conditional probability in (5), observe that for all given $\mathbf{X}_i \in V_\ell^n$ and $\mathbf{X}_j \in V_r^n$, we have, writing $\rho(x) = 1 \wedge x^{-\delta}$,

$$\tilde{\mathbb{P}}(E_{i,j} = 0 \mid \mathbf{X}_i, \mathbf{X}_j) \leq \exp\left(-\rho(\beta^{-1}(T_i \wedge T_j)^\gamma (T_i \vee T_j)^\alpha 5n)\right).$$

Therefore, by writing F_i^n for the empirical distributions of the vertex marks in V_i^n, for $i = \ell, r$, we infer

$$\tilde{\mathbb{P}}(V_\ell^n \not\sim V_r^n) = \tilde{\mathbb{E}}\left[\prod_{i=-n-1}^{-2n} \prod_{j=n}^{2n-1} \tilde{\mathbb{P}}(E_{i,j} = 0 \mid \mathcal{X}_0)\right]$$

$$\leq \tilde{\mathbb{E}}\left[\exp\left(-\sum_{i=-n-1}^{-2n} \sum_{j=n}^{2n-1} \rho(\beta^{-1}(T_i \wedge T_j)^\gamma (T_i \vee T_j)^\alpha 5n)\right)\right]$$

$$= \tilde{\mathbb{E}}\left[\exp\left(\int_0^1 F_\ell^n(\mathrm{d}t) \int_0^1 F_r^n(\mathrm{d}s) \, \rho(\beta^{-1}(t \wedge s)^\gamma (t \vee s)^\alpha 5n)\right)\right].$$

By μ-regularity, we get by [17, Eq. (8)] that $nF_i^n(t) \geq \frac{n}{3}(t - n^{\mu-1})$ and therefore by a change of variables in the last integral we derive for some constant $C > 0$

$$\tilde{\mathbb{P}}(V_\ell^n \not\sim V_r^n) \leq \exp\left(- Cn^2 \int_{n^{\mu-1}}^{1-n^{\mu-1}} \mathrm{d}t \int_t^{1-n^{\mu-1}} \mathrm{d}s \left(1 \wedge \left(\tfrac{5}{\beta} t^\gamma s^\alpha n\right)^{-\delta}\right)\right)$$
$$\leq \exp\left(- Cn^{2-\delta_{\mathrm{eff}}-\varepsilon}\right),$$

where the last step follows by the fact that the order of the integral is driven by the lower integration bound together with the continuity of the integral in μ. This concludes the proof of (i).

The proof of (ii) works similarly. However, the definition of μ-regularity has to be slightly changed. We now say the marks of V_ℓ^n are μ-regular if

(a) $\displaystyle\min_{-2n \leq j \leq -n-1} T_j \geq \lceil n^{-1-\mu}\rceil$ and

(b) $\displaystyle\sum_{j=-n-1}^{-2n} \mathbb{1}_{\{T_j \leq i/\lceil n^{1-\mu}\rceil\}} \leq \frac{2in}{\lceil n^{1-\mu}\rceil}$

which holds with a probability of order $1 - n^{-\mu}$. Using now a lower bound on the distances and performing similar calculations as above yields (ii), cf. [17, Lemma 4.1].

2.3 Existence of a Giant Component

In this section, we use a renormalisation scheme introduced by Duminil-Copin et al. [9] for the existence of an infinite component in one-dimensional long-range percolation on the lattice to construct a component growing linear with a subsequence of $(\mathscr{G}_N^\beta : N > 0)$ from which we derive the existence of a giant component for large enough values of β whenever $\delta_{\mathrm{eff}} < 2$.

We start by defining the scales on which the renormalisation works. For some $K \in \mathbb{N}$, define $(K_n : n \in \mathbb{N})$ by $K_n := (n!)^3 K^n$. Define on each scale the blocks of vertices

$$B_{K_n}^i := \{\mathbf{X}_{K_n(i-1)}, \dots, \mathbf{X}_{K_n i}, \dots, \mathbf{X}_{K_n(i+1)-1}\}, \quad i \in \mathbb{Z},$$

and we abbreviate $B_K = B_K^0$. Each of the scale n blocks consists of $2K_n$ vertices and two consecutive blocks intersect on half of their points. We fix a $\vartheta > 3/4$ and say that the block $B_{K_n}^i$ is ϑ-*good* if it contains a connected component of density at least ϑ. Otherwise, we say the block is ϑ-*bad*. Note that due to the overlapping property, the largest components of two consecutive ϑ-good blocks intersect in at least one vertex.

Consider for some $\vartheta > 3/4$ the sequence $\vartheta_n := \vartheta - 2/(n^3 K)$, where K is chosen large enough to guarantee $\inf_n \vartheta_n > 3/4$. We want to bound the probability of the scale n block B_{K_n} being ϑ_n-bad and we consider the scale $n - 1$ blocks $B_{K_{n-1}}^{-n^3 K-1}, \dots, B_{K_{n-1}}^{n^3 K-1}$ contained in it. If all these blocks are ϑ_{n-1}-good so is B_{K_n} by our choice of $\inf_m \vartheta_m > 3/4$. Therefore, either there must exist at least two disjoint ϑ_{n-1}-bad scale $n-1$ blocks or there is one ϑ_{n-1}-bad block $B^i_{K_{n-1}}$ and all blocks disjoint from it are good. The first event is bounded by

$$\mathbb{P}(\exists \text{ two disjoint } \vartheta_{n-1}\text{-bad blocks}) \leq 2(n^3 K)^2 \mathbb{P}(B_{K_{n-1}} \text{ is } \vartheta_{n-1}\text{-bad})^2. \quad (6)$$

We denote the latter event by A_i and have

$$\mathbb{P}(\{B_{K_n} \text{ is } \vartheta_n\text{-bad}\} \cap A_i)$$

$$\leq \mathbb{P}(B_{K_{n-1}} \text{ is } \vartheta_{n-1}\text{-bad}) \sum_{|i|=0}^{n^3 K - 1} \mathbb{P}(B_{K_n} \text{ is } \vartheta_n\text{-bad} \mid A_i). \quad (7)$$

To calculate the conditional probability, observe that $|i| \notin \{n^3 K - 2, n^3 K - 1\}$ since otherwise B_{K_n} would be ϑ_n-good by our choice of $\vartheta_n = \vartheta_{n-1} - 2/(n^3 K)$. Hence, by the overlapping property, there exists a connected component $\mathscr{C}^i_\ell$ left of the bad block and a connected component $\mathscr{C}^i_r$ on the right, both of density at least ϑ_{n-1}. Further, if both these clusters are connected by an edge, the whole block B_{K_n} again is ϑ_n-good. Hence,

$$\sum_{|i|=0}^{n^3 K - 1} \mathbb{P}(B_{K_n} \text{ is } \vartheta_n\text{-bad} \mid A_i) \leq \sum_{|i|=0}^{n^3 K - 3} \mathbb{P}(\mathscr{C}^i_\ell \nsim \mathscr{C}^i_r \mid A_i).$$

Let $\vartheta^* := \inf_m \vartheta_m (> 3/4)$ and define the 'leftmost' and 'rightmost' vertices of B_{K_n} by

$$V^n_\ell(\vartheta^*) := \{\mathbf{X}_{-K_n}, \ldots, \mathbf{X}_{-K_n + \lfloor \vartheta^* K_{n-1} \rfloor - 1}\} \quad \text{and}$$

$$V^n_r(\vartheta^*) := \{\mathbf{X}_{K_n - \lfloor \vartheta^* K_{n-1} \rfloor}, \ldots, \mathbf{X}_{K_n - 1}\}.$$

We claim $\mathbb{P}(\mathscr{C}^i_\ell \nsim \mathscr{C}^i_r \mid A_i) \leq \mathbb{P}(V^n_\ell(\vartheta^*) \nsim V^n_r(\vartheta^*))$, see Lemma 2 below. Combining this with (6) and (7), we infer for all $n \geq 2$

$$\mathbb{P}(B_{K_n} \text{ is } \vartheta_n\text{-bad})$$

$$\leq \mathbb{P}(B_{K_{n-1}} \text{ is } \vartheta_{n-1}\text{-bad})\big(2n^3 K \, \mathbb{P}(V^n_\ell(\vartheta^*) \nsim V^n_r(\vartheta^*))\big)$$
$$+ 2(n^3 K)^2 \, \mathbb{P}(B_{K_{n-1}} \text{ is } \vartheta_{n-1}\text{-bad})^2$$

$$\leq \mathbb{P}(B_{K_{n-1}} \text{ is } \vartheta_{n-1}\text{-bad})\big(2n^3 K \big(\exp(-CK_{n-1}^{2-\delta_{\mathrm{eff}} - \varepsilon}) + O(K_{n-1}^{1-\mu} e^{-K_{n-1}^\mu})\big)\big)$$
$$+ 2(n^3 K)^2 \mathbb{P}(B_{K_{n-1}} \text{ is } \vartheta_{n-1}\text{-bad})^2$$

$$\leq \frac{1}{100} \mathbb{P}(B_{K_{n-1}} \text{ is } \vartheta_{n-1}\text{-bad}) + 2(n^3 K)^2 \mathbb{P}(B_{K_{n-1}} \text{ is } \vartheta_{n-1}\text{-bad})^2$$

$$(8)$$

for sufficiently large K and the right choice of ε and μ using $\delta_{\mathrm{eff}} < 2$. The second inequality follows from Lemma 1(i) and the fact that $K_{n-1}^{\mu-1} \leq K_n^{\mu-1+\varepsilon'}$ for large enough K.

To deal with the first scale, we condition on $|X_{-K} - X_{K-1}| < 5K$ which holds with an error term exponentially small in K. We choose $\beta > 5K$ so that for all pairs of vertices $\mathbf{X}, \mathbf{Y}$ of B_K, on this event, we have $\beta^{-1}(T_x \wedge T_y)^\gamma (T_x \vee T_y)^\alpha |X - Y| < 1$ and therefore both vertices are connected by an edge by connection rule (2). Hence, on this event, the subgraph B_K is complete. Combining this with (8), we infer inductively for all n and large enough K

$$\mathbb{P}(B_{K_n} \text{ is } \vartheta_n\text{-bad}) \leq \frac{1}{400(n^3 K)^2} \leq \frac{1}{2}.$$

Now observe, that B_{K_n} is contained in the interval $(-2K_n, 2K_n)$ with an error term exponentially small in K_n and therefore uniformly

$$\mathbb{P}(\sharp \mathscr{C}(\mathscr{G}_{2K_n}^\beta) \geq \tfrac{3}{8} K_n) \geq \frac{3}{8}$$

since $\vartheta_n \geq \vartheta^* > 3/4$. We have hence shown that for $\delta_{\text{eff}} < 2$ and large enough K and β, the largest connected component of the subsequence $(\mathscr{G}_{2K_n}^\beta : n \in \mathbb{N})$ grows linearly in time. The existence of a giant component for the whole sequence $(\mathscr{G}_N^\beta : N \in \mathbb{N})$ is then simply a consequence of the ergodicity in our model, cf. [17, Corollary 2.6].

It remains to prove the lemma used in the bound of (8).

Lemma 2. *For all* $|i| \in \{0, \dots, n^3 K - 3\}$, *we have*

$$\mathbb{P}(\mathscr{C}_\ell^i \not\sim \mathscr{C}_r^i \mid A_i) \leq \mathbb{P}(V_\ell^n(\vartheta^*) \not\sim V_r^n(\vartheta^*)).$$

Proof. To shorten notation, we abbreviate $V_\ell = V_\ell^n(\vartheta^*)$ and $V_r = V_r^n(\vartheta^*)$. The proof is based on the idea that belonging to the largest clusters in good boxes gives negative information for not being connected compared to the uniform (i.e. independently sampled) case. Observe first that on the event A_i, we have $\sharp \mathscr{C}_\ell^i \geq \sharp V_\ell$ and $\sharp \mathscr{C}_r^i \geq \sharp V_r$. Let I_ℓ be the (random) set of all indices belonging to the vertices of $\mathscr{C}_\ell^i$ ordered from smallest absolute value to largest and let I_r be the same set for the indices of $\mathscr{C}_r^i$. Let further be $\mathcal{J}_\ell$ a set of $\sharp V_r$-many indices chosen independently from everything else and uniformly among all indices of vertices in B_{K_n} left of the block $B_{K_{n-1}}^i$ and $\mathcal{J}_r$ be the same but for the indices on the right side. Note that due to our construction, the indices are deterministically given. To bound the probability of $\mathscr{C}_\ell^i$ and $\mathscr{C}_r^i$ not being connected by an edge, we first choose a subset of smaller size $\sharp V_\ell = \sharp V_r$ uniformly from both clusters and only ask that there is no edge connecting these. Note that choosing uniform a subset of size $\sharp V_\ell$ from I_ℓ is the same as using the indices in $\mathcal{J}_\ell$, conditioned on $\mathcal{J}_\ell \subset I_\ell$. We infer

$$\mathbb{P}(\mathscr{C}_\ell^i \not\sim \mathscr{C}_r^i \mid A_i)$$

$$= \mathbb{P}\Big(\bigcap_{i_\ell \in I_\ell} \bigcap_{i_r \in I_r} \{E_{i_\ell, i_r} = 0\} \,\Big|\, A_i \Big)$$

$$\leq \frac{\mathbb{P}\Big(\{\mathcal{J}_\ell \subset I_\ell\} \cap \{\mathcal{J}_r \subset I_r\} \cap \big(\bigcap_{i_\ell \in \mathcal{J}_\ell} \bigcap_{i_r \in \mathcal{J}_r} \{E_{i_\ell, i_r} = 0\} \big) \,\Big|\, A_i \Big)}{\mathbb{P}(\{\mathcal{J}_\ell \subset I_\ell\} \cap \{\mathcal{J}_r \subset I_r\} \mid A_i)}.$$

Since $\vartheta^* > 3/4$, on the event A_i, the clusters $\mathscr{C}_\ell^i$ and $\mathscr{C}_r^i$ are the unique largest clusters on the left and right of the block $B_{K_{n-1}}^i$. Therefore, $\{\mathcal{J}_\ell \subset I_\ell\} \cap \{\mathcal{J}_r \subset I_r\}$

is an increasing event, since strengthening the vertices or adding more edges increases the clusters. Note that adding additional vertices to $\mathcal{X}_0$ is equivalent to bringing the vertices closer together which may then also lead to additional edges. Conversely, the event $\{E_{i,j} = 0\}$ is a decreasing event in the sense that $-\mathbb{1}_{\{E_{i,j}=0\}}$ is increasing. Therefore, the FKG-inequality (4) yields

$$\mathbb{P}(\mathscr{C}_\ell^i \not\sim \mathscr{C}_r^i \mid A_i) \leq \mathbb{P}\Big(\bigcap_{i_\ell \in \mathcal{J}_\ell} \bigcap_{i_r \in \mathcal{J}_r} \{E_{i_\ell, i_r} = 0\} \,\Big|\, A_i \Big).$$

Moreover, the existence of edges between vertices outside $B_{K_n}^i$ does not depend on the vertices and edges within this block. Denoting by $\tilde{A}_i$ the *increasing* event that all blocks disjoint from $B_{K_n}^i$ are good, another application of the FKG-inequality yields

$$\mathbb{P}(\mathscr{C}_\ell^i \not\sim \mathscr{C}_r^i \mid A_i) \leq \mathbb{P}\Big(\bigcap_{i_\ell \in \mathcal{J}_\ell} \bigcap_{i_r \in \mathcal{J}_r} \{E_{i_\ell, i_r} = 0\} \,\Big|\, \tilde{A}_i \Big) \leq \mathbb{P}\Big(\bigcap_{i_\ell \in \mathcal{J}_\ell} \bigcap_{i_r \in \mathcal{J}_r} \{E_{i_\ell, i_r} = 0\} \Big).$$

The vertices on the right-hand side are now chosen uniformly at random from all vertices on the left resp. on the right of the bad block. The proof finishes with the observation that in each such sample all vertices and edges have independent and identically distributed marks so that the probability is increased when choosing the left-most resp. right-most vertices maximising the distances between the involved vertices.

2.4   Absence of an Infinite Component

The proof of non-existence of an infinite component for all β when $\delta_{\mathrm{eff}} > 2$ relies on an edge counting argument. We say an edge *crosses the origin* if it connects a vertex left of the origin with one of the right. Here, without loss of generality, we consider $\mathbf{X}_0$ being right of the origin. We show that with a positive probability no such crossing exists. By ergodicity this then holds true for edges crossing any natural number and each component must be finite.

Define for each $n \in \mathbb{N}$ the disjoint sets

$$\Gamma_n^\ell := \{\mathbf{X}_{-2^n}, \dots, \mathbf{X}_{-1}\}, \qquad \Gamma_n^{\ell\ell} := \{\mathbf{X}_{-2^{n+1}}, \dots, \mathbf{X}_{-2^n-1}\},$$
$$\Gamma_n^r := \{\mathbf{X}_0, \dots, \mathbf{X}_{2^n-1}\}, \quad \Gamma_n^{rr} := \{\mathbf{X}_{2^n}, \dots, \mathbf{X}_{2^{n+1}-1}\}.$$

We say a crossing of the origin occurs at stage

$n = 1$, if any edge connects the set $\Gamma_1^\ell \cup \Gamma_1^{\ell\ell}$ and $\Gamma_1^r \cup \Gamma_1^{rr}$ or at stage $n \geq 2$, if any edge connects either Γ_n^ℓ and Γ_n^{rr} or $\Gamma_n^{\ell\ell}$ and Γ_n^{rr} or Γ_n^ℓ and Γ_n^{rr}.
Note that any edges connecting Γ_n^ℓ and Γ_n^r have already been considered at an earlier stage.

Let $\chi(n) \in \{0, 1\}$ denote the indicator of the event that there is a crossing occurring at stage n. Since the events $\{\chi(n) = 0\}$ are all decreasing and thus positively correlated, we have

$$\mathbb{P}\Big(\bigcap_{n \in \mathbb{N}} \{\chi(n) = 0\} \Big) \geq \prod_{n \in \mathbb{N}} \mathbb{P}(\chi(n) = 0).$$

Since the product of the right-hand-side is strictly larger than zero if and only if the sum of the probabilities of the complementary events $\{\chi(n) = 1\}$ converges, the proof finishes by applying Lemma 1(ii) and $\delta_{\text{eff}} > 2$ to

$$\sum_{n \in \mathbb{N}} \mathbb{P}(\chi(n) = 1) \le \mathbb{P}(\chi(1) = 1) + 3 \sum_{n \ge 2} \mathbb{P}(\Gamma_n^{\ell\ell} \sim \Gamma_n^r).$$

Acknowledgement. We gratefully received support by Deutsche Forschungsgemeinschaft (DFG, German Research Foundation) - grant no. 443916008 (SPP 2265) and by the Leibniz Association within the Leibniz Junior Research Group on *Probabilistic Methods for Dynamic Communication Networks* as part of the Leibniz Competition. We would also like to thank Arne Grauer who provided us the R-code we used for Fig. 2.

References

1. Aizenman, M., Newman, C.M.: Discontinuity of the percolation density in one-dimensional $1/|x - y|^2$ percolation models. Comm. Math. Phys. **107**(4), 611–647 (1986). http://projecteuclid.org/euclid.cmp/1104116233
2. Bode, M., Fountoulakis, N., Müller, T.: On the largest component of a hyperbolic model of complex networks. Electron. J. Combin. **22**(3), Paper 3.24, 46 (2015). https://doi.org/10.37236/4958
3. Bringmann, K., Keusch, R., Lengler, J.: Sampling geometric inhomogeneous random graphs in linear time. In: 25th European Symposium on Algorithms, LIPIcs. Leibniz International Proceedings in Informatics, vol. 87, pp. Art. No. 20, 15. Schloss Dagstuhl. Leibniz-Zent. Inform. Wadern (2017)
4. Bringmann, K., Keusch, R., Lengler, J.: Geometric inhomogeneous random graphs. Theoret. Comput. Sci. **760**, 35–54 (2019). https://doi.org/10.1016/j.tcs.2018.08.014
5. Chung, F., Lu, L.: Complex graphs and networks, CBMS Regional Conference Series in Mathematics, vol. 107. Published for the Conference Board of the Mathematical Sciences, Washington, DC; by the American Mathematical Society, Providence, RI (2006). https://doi.org/10.1090/cbms/107
6. Deijfen, M., van der Hofstad, R., Hooghiemstra, G.: Scale-free percolation. Ann. Inst. Henri Poincaré Probab. Stat. **49**(3), 817–838 (2013). https://doi.org/10.1214/12-AIHP480
7. Deprez, P., Wüthrich, M.V.: Scale-free percolation in continuum space. Commun. Math. Stat. **7**(3), 269–308 (2018). https://doi.org/10.1007/s40304-018-0142-0
8. Dorogovtsev, S.N.: Lectures on Complex Networks. Oxford Master Series in Physics, vol. 20. Oxford University Press, Oxford (2010). https://doi.org/10.1093/acprof:oso/9780199548927.001.0001. Oxford Master Series in Statistical Computational, and Theoretical Physics
9. Duminil-Copin, H., Garban, C., Tassion, V.: Long-range models in 1D revisited (2020)
10. Fortuin, C.M., Kasteleyn, P.W., Ginibre, J.: Correlation inequalities on some partially ordered sets. Comm. Math. Phys. **22**, 89–103 (1971). http://projecteuclid.org/euclid.cmp/1103857443
11. Gilbert, E.N.: Random plane networks. J. Soc. Indust. Appl. Math. **9**, 533–543 (1961)

12. Gouéré, J.B.: Subcritical regimes in the Poisson Boolean model of continuum percolation. Ann. Probab. **36**(4), 1209–1220 (2008). https://doi.org/10.1214/07-AOP352
13. Gracar, P., Grauer, A., Lüchtrath, L., Mörters, P.: The age-dependent random connection model. Queueing Syst. **93**(3-4), 309–331 (2019). https://doi.org/10.1007/s11134-019-09625-y
14. Gracar, P., Grauer, A., Mörters, P.: Chemical distance in geometric random graphs with long edges and scale-free degree distribution. Comm. Math. Phys. **395**(2), 859–906 (2022). https://doi.org/10.1007/s00220-022-04445-3
15. Gracar, P., Heydenreich, M., Mönch, C., Mörters, P.: Recurrence versus transience for weight-dependent random connection models. Electron. J. Probab. **27**, 1–31 (2022). https://doi.org/10.1214/22-EJP748
16. Gracar, P., Lüchtrath, L., Mörters, P.: Percolation phase transition in weight-dependent random connection models. Adv. Appl. Probab. **53**(4), 1090–1114 (2021). https://doi.org/10.1017/apr.2021.13
17. Gracar, P., Lüchtrath, L., Mönch, C.: Finiteness of the percolation threshold for inhomogeneous long-range models in one dimension (2022). https://doi.org/10.48550/ARXIV.2203.11966. https://arxiv.org/abs/2203.11966
18. Hall, P.: On continuum percolation. Ann. Probab. **13**(4), 1250–1266 (1985). http://links.jstor.org/sici?sici=0091-1798(198511)13:4⟨1250:OCP⟩2.0.CO;2-U&origin=MSN
19. Heydenreich, M., van der Hofstad, R., Last, G., Matzke, K.: Lace expansion and mean-field behavior for the random connection model (2020)
20. Hirsch, C.: From heavy-tailed Boolean models to scale-free Gilbert graphs. Braz. J. Probab. Stat. **31**(1), 111–143 (2017). https://doi.org/10.1214/15-BJPS305
21. van der Hofstad, R.: The giant in random graphs is almost local (2021). https://doi.org/10.48550/ARXIV.2103.11733. https://arxiv.org/abs/2103.11733
22. van der Hofstad, R., van der Hoorn, P., Maitra, N.: Local limits of spatial inhomogeneous random graphs (2021). https://doi.org/10.48550/ARXIV.2107.08733. https://arxiv.org/abs/2107.08733
23. van der Hofstad, R., van der Hoorn, P., Maitra, N.: Scaling of the clustering function in spatial inhomogeneous random graphs (2022). https://doi.org/10.48550/ARXIV.2212.12885. https://arxiv.org/abs/2212.12885
24. Jahnel, B., Lüchtrath, L.: Existence of subcritical percolation phases for generalised weight-dependent random connection models (2023). https://doi.org/10.48550/ARXIV.2302.05396. https://arxiv.org/abs/2302.05396
25. Komjáthy, J., Lapinskas, J., Lengler, J.: Penalising transmission to hubs in scale-free spatial random graphs. Ann. Inst. Henri Poincaré Probab. Stat. **57**(4), 1968–2016 (2021). https://doi.org/10.1214/21-AIHP1149
26. Komjáthy, J., Lodewijks, B.: Explosion in weighted hyperbolic random graphs and geometric inhomogeneous random graphs. Stoch. Process. Appl. **130**(3), 1309–1367 (2020). https://doi.org/10.1016/j.spa.2019.04.014. https://www.sciencedirect.com/science/article/pii/S0304414918301601
27. Last, G., Penrose, M.: Lectures on the Poisson Process. Cambridge University Press, Cambridge (2017). https://doi.org/10.1017/9781316104477
28. Lüchtrath, L.: Percolation in weight-dependent random connection models. Ph.D. thesis, Universität zu Köln (2022). https://kups.ub.uni-koeln.de/64064/
29. Meester, R., Penrose, M.D., Sarkar, A.: The random connection model in high dimensions. Stat. Probab. Lett. **35**(2), 145–153 (1997). https://doi.org/10.1016/S0167-7152(97)00008-4. https://www.sciencedirect.com/science/article/pii/S0167715297000084

30. Newman, C.M., Schulman, L.S.: One-dimensional $1/|j-i|^s$ percolation models: the existence of a transition for $s \leq 2$. Comm. Math. Phys. **104**(4), 547–571 (1986). http://projecteuclid.org/euclid.cmp/1104115167

31. Penrose, M.D.: Connectivity of soft random geometric graphs. Ann. Appl. Probab. **26**(2), 986–1028 (2016). https://doi.org/10.1214/15-AAP1110

32. Schulman, L.S.: Long range percolation in one dimension. J. Phys. A **16**(17), L639–L641 (1983). http://stacks.iop.org/0305-4470/16/639

33. Yukich, J.E.: Ultra-small scale-free geometric networks. J. Appl. Probab. **43**(3), 665–677 (2006). https://doi.org/10.1239/jap/1158784937

Unsupervised Framework for Evaluating Structural Node Embeddings of Graphs

Ashkan Dehghan[1], Kinga Siuta[1,2], Agata Skorupka[1,2], Andrei Betlen[3], David Miller[3], Bogumił Kamiński[2], and Paweł Prałat[1(✉)]

[1] Department of Mathematics, Toronto Metropolitan University,
Toronto, ON, Canada
{ashkan.dehghan,pralat}@torontomu.ca
[2] Decision Analysis and Support Unit, SGH Warsaw School of Economics,
Warsaw, Poland
bogumil.kaminski@sgh.waw.pl
[3] Patagona Technologies, Pickering, ON, Canada

Abstract. An embedding is a mapping from a set of nodes of a network into a real vector space. Embeddings can have various aims like capturing the underlying graph topology and structure, node-to-node relationship, or other relevant information about the graph, its subgraphs or nodes themselves. A practical challenge with using embeddings is that there are many available variants to choose from. Selecting a small set of most promising embeddings from the long list of possible options for a given task is challenging and often requires domain expertise. Embeddings can be categorized into two main types: classical embeddings and structural embeddings. Classical embeddings focus on learning both local and global proximity of nodes, while structural embeddings learn information specifically about the local structure of nodes' neighbourhood. For classical node embeddings there exists a framework which helps data scientists to identify (in an unsupervised way) a few embeddings that are worth further investigation. Unfortunately, no such framework exists for structural embeddings. In this paper we propose a framework for unsupervised ranking of structural graph embeddings. The proposed framework, apart from assigning an aggregate quality score for a structural embedding, additionally gives a data scientist insights into properties of this embedding. It produces information which predefined node features the embedding learns, how well it learns them, and which dimensions in the embedded space represent the predefined node features. Using this information the user gets a level of explainability to an otherwise complex black-box embedding algorithm.

Keywords: Node Embeddings · Structural Node Embeddings

1 Introduction

Inspired by early work in word embedding techniques [17], node, edge and graph embedding algorithms have gained a lot attention in the machine learning community, in recent years. Indeed, learning an accurate and useful latent representation from network-data is an important and necessary step for any successful

M. Dewar et al. (Eds.): WAW 2023, LNCS 13894, pp. 36–51, 2023.
https://doi.org/10.1007/978-3-031-32296-9_3

machine learning task, including node classification [18], anomaly detection [3], link prediction [11], and community detection [19] (see also surveys [4,10]).

In this paper, we distinguish two families of node embeddings: classical node embeddings and structural node embeddings. The first family is very rich with already over 100 algorithms proposed in the literature. Informally speaking, *classical node embeddings* fall into a broad and diverse family of embeddings that try to assign vectors in some high dimensional space to nodes of the graph that would allow for its approximate reconstruction using such encapsulated information. Different classical embedding algorithms use different approaches to achieve this task. Some of them, in order to extract useful information from graphs, try to create an embedding in a geometric space by assigning coordinates to each node such that nearby nodes are more likely to share an edge than those far from each other. Some other approaches postulate that pairs of nodes that have overlapping neighbourhoods (not necessarily intermediate ones) should have similar representations in the embedded space. Independently, the techniques to construct the desired classical embeddings can be broadly divided into the following three groups: linear algebra algorithms, random walk based algorithms, and deep learning methods [1,16].

Classical embeddings work well for machine learning tasks such as link prediction. However, as the study of [22] shows, they do not guarantee good performance, for example, in tasks such as community labeling that can be viewed as a classification task or role detection. The reason is that in these challenging-for-classical-embeddings machine learning problems, when doing inference, it is important to preserve structural characteristics of nodes. Informally speaking, by structural characteristics of nodes we mean the structure of nodes' egonets, which is the induced subgraphs of given nodes and their neighbourhoods up to some fixed depth. The simplest form of one-dimensional structural embeddings are node features such as degree or local clustering coefficient. Indeed, node features have been used extensively since the very beginning of network analysis, as most of them have natural interpretations and are usually relatively easy to compute. From the standpoint of this discussion, it is important to highlight that such node features do not depend on node labels, but rather on the relationships between them. For example, two nodes might both have large and comparable degrees or similar pageranks (and, as a result, end up close to each other in the embedded space) but be distant from each other in terms of concrete neighbours (and so they would be far apart in classical embeddings). The already mentioned study [22] shows that such node features are efficient in various tasks such as community labeling. The reason is that often the role of a node within a graph is an important predictor of some features but not necessarily its concrete neighbours. Since using hand-crafted node features in various machine learning tools has proven to be a useful technique, researcher have developed various *structural embedding algorithms* such as **RolX** [12], **Struct2Vec** [21], **GraphWave** [7] and **Role2Vec** [2]. Such embedding algorithms try to capture structural characteristics of nodes, that is, put nodes that have similar structural characteristics close together in the embedded space. Again, like with classical embeddings, implementations of structural embeddings differ in the way how they define similarity between neighbourhoods of two nodes.

There are two important questions to consider when studying embedding algorithms. The first question is concerned with what node-features about the graph is learned by a given embedding algorithm. And the second focuses on how well a given node feature is learned by the set algorithm. Answering and understanding these questions is crucial for practitioners of the field, as it will dictate which embedding algorithm is optimal for a given task. Of course, the decision of which algorithm to use might also depend on the properties of the investigated network [6]. There are existing works by [13–15], which aim to answer these questions for the classical type embedding algorithms[1]. There is however no such work to our knowledge that has been done to answer these questions for structural embedding algorithms. In this work, we introduce an unsupervised technique for quantifying how well a given embedding algorithm learns a predefined set of structural features. This provides an explainability of the embedding space in terms of structural-node-features, in addition to allowing one to compare between various different algorithms to identify the most optimal embedding for a given application.

2   Framework

2.1   Input/Output

In this section we introduce our framework and highlight its properties. The goal of the framework is to evaluate possible correlations of various node embeddings with a number of classical node features of a single graph $G = (V, E)$ on $n = |V|$ nodes. The input consists of

- k dimensional node embedding—k vectors of real numbers, each of length n,
- ℓ node features—ℓ vectors of real numbers, each of length n.

The framework outputs the following

- a real number (represented by symbol ψ) from the interval $[0, 1]$ representing how well given feature vectors may be approximated by given embedding vectors; $\psi = 0$ indicates a good approximation and the other extreme value, $\psi = 1$, represents a bad approximation; both pre- and post-optimization values of ψ are returned, where the post-optimization ψ value is computed by minimizing ψ as a function of a vector $\mathbf{w}$—formal definition and more details will be provided soon,
- a vector $\mathbf{w}$ of non-negative real values of length k and L^1-norm equal to 1 that indicates which embedding dimensions contribute to the explanation of features; here, larger values correspond to larger contribution; the $\mathbf{w}$ vector consists of the weights in the embedding distance computation, and is used to identify which embedding dimension the structural feature is mapped onto.

---

[1] https://github.com/KrainskiL/CGE.jl.

The structure of our framework is designed to output a quantitative metric ψ, which measures how well an embedding algorithm has learned a given feature (or a collection of features). This metric can be used to both identify what features embedding algorithms learn, in addition to how well they learn those features. A more comprehensive explanation of this is given in the following section.

2.2 Formal Description of the Algorithm

In our framework, nodes are clustered (using k-means clustering) in the feature-space, and distance between sampled nodes in the feature-space are calculated and compared to the distance measured in the embedded-space. Therefore, the algorithm has a few parameters that the user might experiment with but each of them has a default value:

- s: the number of clusters in the feature space generated by the k-means algorithm (by default, $s = \sqrt{n}$, where n is the number of nodes of a network); the value $s = \sqrt{n}$ is a safe estimated to ensure the convergence and stability of the calculated ψ metric—more on this is discussed in the following section,
- p: the fraction of sampled pairs of nodes that are from the same cluster (by default, $p = 0.5$),
- c: the total number of sampled pairs of nodes (by default, $c = \min\{10^5, n^2/s\}$; apart from a natural upper bound of 10^5, for small networks we need to make sure that the number of pairs of nodes sampled within clusters, $p \cdot c$, is at most the number of all pairs of nodes from the same cluster; indeed, at the worst case scenario each cluster could consist of n/s nodes and so there could be only $\binom{n/s}{2} \cdot s \approx n^2/(2s)$ pairs of nodes within clusters; this would cause a problem as the algorithm samples pairs without replacement),
- standardization method: we provide two methods, MinMax that scales and translates each feature individually such that all of them are in the range between zero and one, and StandardScaler that scales features such that the mean and the standard deviation are equal to zero and, respectively, one (by default, we use the StandardScaler normalization).

The algorithm performs the following steps:

1. **Standardization.** Transform all feature and embedding vectors using one of the two methods, MinMax or StandardScaler. After this transformation, all vectors are appropriately normalized and standardized. As a result, later steps are invariant with respect to any affine transformation of these vectors.
2. **Clustering.** Perform the classical k-means clustering of nodes (into s clusters) in the feature space using the selected metrics. Let $(c_1, \ldots, c_s)$ with $n = \sum_{i=1}^{s} c_i$ be the distribution of cluster sizes.
3. **Sampling.** There are two types of pairs of nodes that are independently sampled as follows.

a) sample

$$\hat{m} = \min\left\{\left\lfloor p \cdot c\right\rfloor, \sum_{1 \le a \le s} \binom{c_a}{2}\right\}$$

unique pairs of nodes within clusters; a single pair of nodes is sampled by first selecting cluster i of size c_i with probability equal to

$$p(i) = \frac{\binom{c_i}{2} - x_i}{\sum_{1 \le a \le s}\left(\binom{c_a}{2} - x_a\right)},$$

where x_i is the number of pairs already sampled from cluster i, and then selecting a pair of nodes from the chosen cluster, uniformly at random; if a pair of nodes sampled this way is already present in the sampled set we discard it, otherwise we keep it.

b) sample

$$\bar{m} = \min\left\{\left\lfloor (1 - p) \cdot c\right\rfloor, \sum_{1 \le a < b \le s} c_a c_b\right\}$$

unique pairs of nodes that are between clusters; a single pair of nodes is sampled by first selecting two clusters i, j $(i < j)$ with probability equal to

$$p(i, j) = \frac{c_i c_j - x_{i,j}}{\sum_{1 \le a < b \le s}(c_a c_b - x_{a,b})},$$

where $x_{i,j}$ is the number of pairs between cluster i and cluster j already sampled, and then selecting one node from each of the chosen clusters, uniformly at random; if a pair of nodes sampled this way is already present in the sampled set we discard it, otherwise we keep it.

4. **Computing Feature Distance.** For each of the sampled pairs of nodes, compute the corresponding distance in the ℓ-dimensional feature space d_f. For the Euclidean metric we have

$$d_f(v_i, v_j) = \sqrt{\sum_{1 \le a \le \ell} (f_a^i - f_a^j)^2},$$

where $(f_1^i, \ldots, f_\ell^i)$ and $(f_1^j, \ldots, f_\ell^j)$ are features of nodes v_i and, respectively, v_j.

5. **Computing Embedded Distance.** Suppose for a moment that a normalized vector of non-negative weights $\mathbf{w} = (w_1, \ldots, w_k)$ with $\sum_{i=1}^{k} w_i = 1$ is fixed. For each of the sampled pairs of nodes, compute the corresponding distance in the k-dimensional embedded space d_e. The weighted Euclidean distance is given by

$$d_e(v_i, v_j) = \sqrt{\sum_{1 \le a \le k} w_a (e_a^i - e_a^j)^2},$$

where $(e_1^i, \ldots, e_k^i)$ and $(e_1^j, \ldots, e_k^j)$ are embeddings of nodes v_i and, respectively, v_j.

6. **Correlation between the two spaces**. To compute the correlation between the two spaces, we define a metric $\psi = 1 - r^2 \in [0,1]$, where $r \in [-1,1]$ is the Pearson correlation between vectors in the embedding space and vectors in the feature space. As a result, ψ is defined such that both large positive (close to 1) and large negative (close to -1) correlation would have small values (close to 0). This is done so that the optimization scheme (see the next bullet-point) is more stable.

7. **Optimization**. Optimize vector $\mathbf{w}$ to minimize ψ, where the final value of ψ is referred to as the post-optimization ψ. These optimized vectors reflect the importance of embedded dimensions for selected features. We note that the optimization is done using Quasi-Newtonian bounded constraint minimization technique from *Scipy* Optimize method. We note that the pre-optimization ψ value measures the overall raw embedding of a particular feature. To measure how well a feature is learned by a particular embedding algorithm, ψ is optimized against that feature. The optimization process removes (or minimizes) any embedded information that does not contribute to the representation of the feature at study. Therefore, we use the post-optimization ψ value to conduct all experiments in this study.

2.3   Properties

Let us briefly highlight some basic and desired properties of the framework which, in particular, justify its design and show its potential usefulness.

- The framework is designed in such a way that affine transformations of any of the feature or embedding vectors do not change the results.
- The framework does not assume any particular type of the relationship between feature space and embedded space. Instead, it is desired that if two nodes are close in the feature space, then they are also close in the embedded space (with a proper metrics/weighting in the embedded space).
- The sampling strategy used in the framework has the following consequence. Achieving a good ψ score ensures that close pairs of nodes in the feature space are close in the embedded space. On the other hand, if some pairs of points are far in the feature space, then the framework puts less weight on the fact whether they are close or not in the embedded space. The rationale behind this property is that a typical pair of nodes are likely to be far in both spaces and so the framework should not pay too much attention to these pairs.
- An embedding algorithm might learn many node features, which may not contribute to the representation of particular structural feature. This additional learned information can be removed and minimized by adjusting the weights associated with appropriate embedding dimensions. For example, the feature *PageRank* may get mapped to dimension of 1 (out of 8) of an embedding space. In this case, dimensions 2 to 7 do not contribute to the representation of *PageRank* and can be removed by setting the weights for those dimensions to 0. This process is done automatically by our framework during the optimization process of the ψ value.

3   Experimentation

In this section, we investigate and analyze various desired algorithmic properties of our framework. We focus on six embedding algorithms, four structural ones (**LSME** [5], **Role2Vec** [2], **Struc2Vec** [21], and **RolX** [12]) and two classical ones (**Node2Vec** [9] and **DeepWalk** [20]). The goal of our analysis here is to understand and analyze various properties of our framework. In addition, we showcase how one could use our framework in applications such as node classification, by investigating the performance of a number of node and structural embedding algorithms. We break up our analysis into two main parts. First (Subsect. 3.2), we explore some of the basic properties of our framework, such as algorithm stability and behaviour. Second (Subsect. 3.3), we showcase the application of our framework in a node classification case study. In this section, we use the default hyper-parameters for every embedding algorithm.

3.1   Synthetic Graphs Design

For experiments in this section we use synthetically generated graph $\mathcal{G}$ which is composed of three structurally distinct sets of subgraphs. As shown in Fig. 1, these subgraphs are labelled Web, Star and dStar. The Web and Star subgraphs each have three types of nodes (w0, w1 and w2) and (s0, s1, s2), while dStar subgraph has two types of nodes (ds0, ds1). The overall synthetic graph is created by joining N_w Web, N_s Star and N_{ds} dStar subgraphs by randomly creating links between w2, s2 and ds1 nodes. The edge creation process is as follows; from joined set of w2, s2 and ds1 nodes randomly select two nodes n_a and n_b. If $n_a \neq n_b$ and $e_{ab} \notin E$, where E is the set of edges of $\mathcal{G}$, then create an edge e_{ab}. Repeat this process until all w2, s2 and ds1 nodes are connected to at least one other node. Based on this description, we can fully define our synthetic graph using 8 parameters: $\mathcal{G}(\{N_w, N_s, N_{ds}\}, \{k_{w1}, k_{w2}\}, \{k_{s1}, k_{s2}\}, \{k_{ds1}\})$. Here, N_w, N_s, N_{ds} are the number of Web, Star and dStar subgraphs in the overall graph and, k_x correspond to the number of nodes in layer x of each subgraph. Each layer of the subgraphs is connected to the previous/next layers as shown in Fig. 1. For example, $k_{w1} = 5$, based on Fig. 1. In this section, we create a synthetic graph with the following parameters: $\mathcal{G}(\{N_w = 200, N_s = 200, N_{ds} = 200\}, \{k_{w1} = 5, k_{w2} = 10\}, \{k_{s1} = 5, k_{s2} = 10\}, \{k_{ds1} = 5\})$, resulting a synthetic graph with 7,600 nodes. We have chosen this structure for our synthetic graph to allow for a simple yet structurally distinct nodes to be used for our classification tasks. Since our framework is designed for structural embedding algorithms, we wanted to use synthetic graphs where nodes have known structural roles (ground-truth). As we shall show in Sect. 3.3, we use the synthetic graph described above to build classifiers for identifying root nodes S_0, and analyze each embedding algorithm's performance for this task.

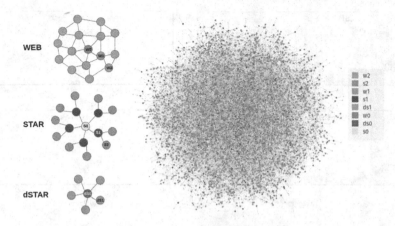

Fig. 1. Synthetic graph $\mathcal{G}(\{N_w = 200, N_s = 200, N_{ds} = 200\}, \{k_{w1} = 5, k_{w2} = 10\}, \{k_{s1} = 5, k_{s2} = 10\}, \{k_{ds1} = 5\})$ composed of a collection of Web, Star and dStar subgraphs.

3.2   Algorithmic Properties of the Framework

In this section, we analyze various algorithmic properties of our framework such as the convergence and stability of various metrics and the behaviour of structural vs. classical embedding algorithms. As we described in Sect. 2.2, the quality of learned representation of a structural feature (for example degree centrality) is measured using the post-optimization ψ value. The optimization is done by minimizing ψ as a function of weights associated with each embedding dimension. To test the effectiveness of this approach, we performed two experiments. In each experiment, we embed the synthetically created graph described in the previous section using either **LSME** or a fixed-embedding algorithm. Here, the fixed-embedding maps the simplest centrality measure, degree centrality, directly onto one of the N embedding dimensions. Furthermore, the other $N - 1$ dimensions are filled with random numbers. This simulates a synthetic embedding algorithm, which learns a perfect representation of a feature and maps it onto one of the dimensions of the embedding space. For our experiments, we used $N = 8$ as the dimension of both embedding algorithms. Once the embedding vectors are generated, we use our framework to measure the performance of each embedding with respect to the degree centrality. In other words, we measure how well did each embedding learn the representation of degree centrality. The purpose for this experiment is to showcase the optimization process and highlight how our framework can be used to study how well features are mapped into the embedded space.

Figure 2 shows the results for the pre and post optimization values for the weights associated with each embedding dimension. In the pre-optimization state (top-left and bottom-left), the values for the weights are set randomly. The optimization algorithm then identifies the dimensions which the representation of the

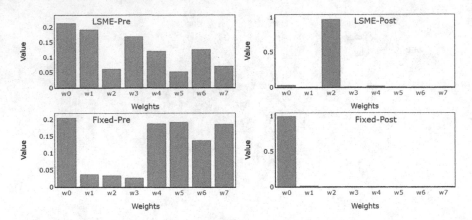

Fig. 2. Pre and post optimization values for weights associated with each embedding dimension (8 dimensional embedding). Top-left and bottom-left figures show preoptimization random initialization of the weights for **LSME** embedding and fixed embedding, respectively. Top-right and bottom-right are the post-optimization values of the weight for **LSME** and fixed embedding.

degree centrality was mapped onto. As expected, the post-optimization weights for the fixed embedding is collapsed to only dimension 0 (w0), which holds a copy of degree centrality for the nodes. For the **LSME** algorithm, degree centrality was mapped by the embedding algorithm onto primarily dimension 2 and partially onto dimension 0 and 4. It is important to note that the post optimization weights by themselves are not complete measure of how well the embedding has learned the node feature. To capture the complete picture, one has to also consider the post optimization score ψ—more on this soon. The experiments in Fig. 2 were repeated multiple times with randomized initial weights, and the results of the experiments were consistent with the above findings.

Before we dive deeper into the other properties of the framework, we want to consider the stability of the algorithm as a function of the node sample size (parameter c) and the number of clusters produced by k-means algorithm (parameter s). To apply the framework to large graphs, we would want our algorithm to converge for both $c \ll n$ and $s \ll n$, where n is the number of nodes of the graph. To measure the stability of the algorithm, we perform two experiments using **LSME** and **Role2Vec** on a synthetic graph. The performance of the embedding algorithms are measured using the degree centrality as the node feature. Figures 3 and 4 show the convergence of ψ as a function of the normalized number of clusters and the normalized sample size respectively, where normalization is done with respect to the size of the graph. Let us first consider the behaviour of ψ as a function of s, the number of clusters. It is important to note that while we vary s in this experiment, values for other parameters are kept as default. As we can see in Fig. 3, ψ converges to its long-run average when the number of clusters is approximately 2% to 3% of the total size of the graph. Both here and in Fig. 4, the long-run average is defined as the expected value for

ψ as sample size or number of clusters approaches the size of the graph. Finally, we conclude that our default value for the number of clusters ($s = \sqrt{n}$) is a good approximation since for the current experiment ($n = 1,000$) the number of clusters is approximately 3% of the size of the graph ($\frac{\sqrt{1,000}}{1,000} \approx 0.03$). Next, we look at the convergence of ψ as a function of c, the sample size. Similarly to the previous experiment, we vary c while setting other parameters to their default values. As one can see in Fig. 4, the value for ψ converges for sample of sizes greater than or equal to 20% to 30% of the size of the graph. The results of our experiments point at two facts. First that the algorithm converges and is stable for both $c \ll n$ and $s \ll n$. Second, the default values for the hyperparameters are good and safe estimates.

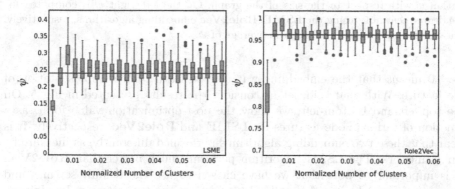

Fig. 3. Post-optimization score ψ as a function of normalized number of clusters. Normalization is with respect to the size of the graph. On the left/right ψ is computed for the degree centrality using the **LSME/Role2Vec** embedding algorithms, respectively. The horizontal lines are the long-run average of ψ.

We now turn our attention to experiments comparing the general behaviour of structural embedding as compared to classical node embedding algorithms. Classical node embedding algorithms such as **Node2Vec** [9] and **DeepWalk** [20] have difficulty learning structural properties of graphs. To showcase that our framework can be used as an unsupervised method for capturing this effect, we perform four experiments using two structural and two classical embedding algorithms. All four algorithms are ran against synthetic graphs (created using the procedure in Sect. 3.1) to generate 8-dimensional embedding vectors. Each embedding is then evaluated using our framework, where its performance is measured against 12 classical and widely used node features. For each node feature, we compute the post optimization ψ value. As we noted previously, the value of ψ is inversely correlated to how well the embedding was able to learn a given representation. In particular, $\psi = 1$ means that the embedding was not able to learn anything for a given feature, and in the other extreme

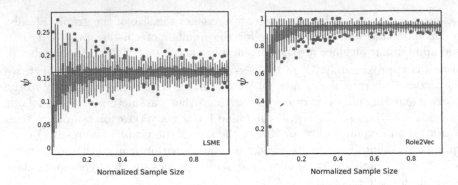

Fig. 4. Post-optimization score ψ as a function of normalized sample size. Normalization is with respect to the size of the graph. On the left/right ψ is computed for the degree centrality using the **LSME**/**Role2Vec** embedding algorithms, respectively. The horizontal lines are the long-run average of ψ.

$\psi = 0$ means that the embedding was able to learn prefect representation of the feature. With that said, let us consider the results presented in Fig. 5. On the top-left and bottom-left, we show the post-optimization value for ψ as a function of various node features for **LSME** and **Role2Vec**, respectively. It is clear that these two embedding algorithms performed differently, as measure by our framework. The **LSME** algorithm performs much better than **Role2Vec**. It is important to note that we have chosen to use the default settings and parameters for each algorithm and did not perform any optimization. In addition, we have only focused on a set of 12 structural features, while algorithms could be learning features not in our set. While the structural embedding algorithm **LSME** was able to learn some structural node features, as expected, classical embedding algorithms (**Node2Vec** and **DeepWalk**) struggle with this task. This is clearly shown in the top-right and bottom right-plots of Fig. 5, where the post-optimization ψ values for all node features are close to 1. Lastly, both **Node2Vec** and **DeepWalk** perform similarly to one another, indicating the similarity in the underlying algorithms.

3.3   Role Classification Case Study

In this subsection, we explore the use case of our framework for analyzing role classification in a synthetic network introduced earlier. A common task in network analysis is to classify nodes based on the role the nodes play in their local network structure. To build features for a classification algorithm, one could either use manually calculated structural properties of the nodes (node features) or leverage structural or node embedding features; as an automated way of learning various structural properties of nodes. One major challenge with using some embedding algorithms as a source for feature engineering, is the lack of explainability of the learned representations. It is not easy to identify what structural properties of the nodes are learned and how a given learned representation is mapped onto the embedding space. To explore these ideas and showcase one possible use case of our framework, we consider the synthetic network introduced in Subsect. 3.1, and use w0, s0 and ds0 root nodes as the target nodes we would like to classify. The root nodes considered here have very similar local structure, creating a relatively challenging tasks for a classifier. The goal of our analysis is to design and build a classifier using both node features and features extracted from various embedding algorithms. Lastly, we show how one could use our framework as an unsupervised technique to gain insight into the performance of embedding algorithms in applications such as role classification. We use six embedding algorithms, two classical algorithms (**Node2Vec** and **DeepWalk**) and four structural based ones (**LSME**, **Struc2Vec**, **RolX**, and **Role2Vec**). We hope to answer the following questions: can one use our framework to identify embedding algorithms that best learn various structural properties of nodes, which could hint at their performance in a role classification task? Additionally, can one extract insights into the predictability strength of each node feature and how those features are learned by a given embedding algorithm?

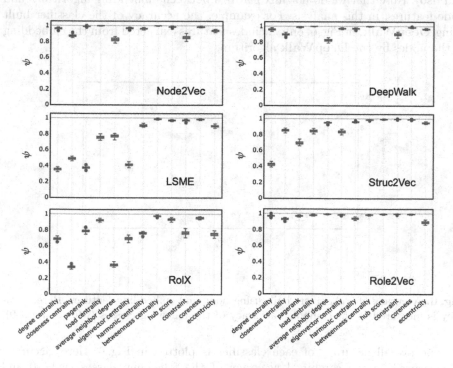

Fig. 5. Post-optimization ψ values (y-axis) computed as a function of 12 node features (x-axis) for various classical and structural embedding algorithms.

We first start by analyzing each embedding algorithm using our framework. We consider 12 node features as a benchmark and compute ψ for each feature. The performance of each embedding algorithm is presented in Fig. 5. As before,

ψ is inversely proportional to how well an embedding algorithm has learned a given feature, where $\psi = 1$ means that a given feature was not learned by the algorithm. As we can see in Fig. 5, classical embedding algorithms (**Node2Vec** and **DeepWalk**) fail to learn the structural properties of the graphs. This aligns with our expectations, since classical algorithms are designed to learn classical node properties. Furthermore, **Role2Vec** also fails to learn any structural proprieties of the graph, while **LSME**, **Struc2Vec**, and **RolX** perform quite well. It is important to note that we used the default hyper-parameters for each algorithm, and it is possible to achieve better results if one optimizes the learning process. Using current results, one would expect **LSME**, **Struc2Vec**, and **RolX** algorithms to perform better than **Node2Vec**, **DeepWalk**, and **Role2Vec** in classification tasks, where the structural properties of the nodes are of importance. With the results from Fig. 5 in mind, we build 7 classifiers with the goal of classifying w0, s0 and ds0 nodes in graph G using features built using manually computed node features and features from each of our six embedding algorithms. The classifiers are trained to predict 3 classes, one for each root node (w0, s0 and ds0). Note that we do not mix features between embedding algorithms and node-features, in this analysis. For example, the accuracy of the classifier built using **DeepWalk**, in Fig. 6, only includes features extracted from the embedding of the nodes by the **DeepWalk** algorithm.

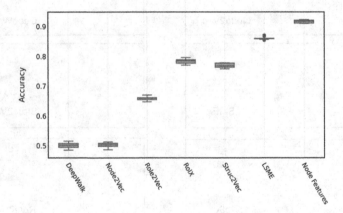

Fig. 6. Accuracy of 7 classifiers built using node and embedding features. Here, accuracy is measured as the combined accuracy of the following classes (w0, s0 and ds0).

The overall accuracy of each classifier is plotted in Fig. 6. Here, accuracy is measured as the combined accuracy of the following classes (w0, s0 and ds0). For each embedding, we train 10 models and select the best performing model and overage the performance across 10 samples. As expected, based on our analysis in Fig. 5, classifiers built using node features, **Struc2Vec**, **RolX**, and **LSME** perform much better than those built using **Role2Vc**, **DeepWalk**, and **Node2Vec**. It is important to consider the following when analyzing these results. The fact that a classifier built using solely node feature performs well, indicates that any embedding that learns structural properties of the node should also perform well. However, this logic does not apply in reverse. The poor performance of an embedding based on our framework does not necessarily indicate that it will perform poorly in a classification task, since there may be features with predictive power which are not captured by the reference features of our framework. We note that, the set of reference features is modifiable and could be updated to include additional structural features. One could extend the 12 features in the benchmarking set to capture high order structural properties of the graph, to allow for a more extensive list of structural properties. Lastly, we point out that one could use the output of the framework to study the specific features learned by each embedding algorithm. For example, both **LSME** and **Struc2Vec** fail to learn *Constrain*, which is the measure of Burt's Constraint [8] for each node, (see Fig. 5) as a structural feature, while **RolX** performs better in this regard. This is an important observation in scenarios where one would want to combine the features from different embedding algorithm to built feature sets with more predictive power. It is natural that each embedding algorithm learns slightly different properties of the graph. Our framework can be used as a tool to map out the embedding space and understand it through the lens of structural features of the graph.

4   Conclusion

In this work, we introduced an unsupervised embedding evaluation framework which can be used to both explain what structural properties of nodes embedding algorithms learn, in addition to how well each algorithm learns a particular structural feature. As we noted above, for tasks such as role-discovery or role-classification, one needs to rely on structural properties of nodes learned by structural embedding algorithms. However, there are numerous challenges with using structural embedding algorithms. First, there is a diverse set of structural features that an algorithm could learn. Therefore, it is not easy to define a single metric for measuring the performance of structural embedding algorithms. Second, measuring performance of embedding algorithms is often done using supervised techniques, which relies on the availability of labeled dataset.

In Sect. 2, we introduced a framework, which addresses the above two challenges. In our framework, we introduce a collection of core structural features, against which one could measure the performance of a structural embedding algorithm. In addition, we introduce a technique for performing these measurements in an unsupervised way, which avoids the need for the availability of labelled datasets. By introducing a mapping between the embedding and the feature space, we are able to define a metric (ψ) for measuring the performance of embedding algorithms. In addition, we can use this metric to explain which features are learned by a given algorithm. As we have shown in Sect. 3.2, this feature of our framework is especially useful for the explainability of algorithms that rely on deep-learning such as **LSME**. Furthermore, using a synthetic graph as a benchmark, we showcased several use cases for our framework.

In Sect. 3.3, we showed that one could use our framework to measure the performance of a number of classical and structural embedding algorithms against a set of structural features. The performance of the embedding algorithms, as measured by ψ, correlates with the performance of the algorithms in a role classification task. This highlights the utility of our framework, which can be used to gain insights into the performance of embedding algorithms in scenarios where labeled data is not available. In addition, one could use our framework to identify difficult to embed structural features and use the ψ value as iterative way of modifying an embedding algorithm to learn specific feature-sets. The unsupervised framework developed and showcased in this work can be used as a versatile and useful tool for practitioners studying structural properties of complex networks.

References

1. Aggarwal, M., Murty, M.N.: Machine Learning in Social Networks: Embedding Nodes, Edges, Communities, and Graphs. Springer, Singapore (2021). https://doi.org/10.1007/978-981-33-4022-0
2. Ahmed, N.K., et al.: Learning role-based graph embeddings. arXiv preprint arXiv:1802.02896 (2018)
3. Akoglu, L., Tong, H., Koutra, D.: Graph based anomaly detection and description: a survey. Data Min. Knowl. Disc. **29**(3), 626–688 (2015)
4. Chami, I., Abu-El-Haija, S., Perozzi, B., Ré, C., Murphy, K.: Machine learning on graphs: a model and comprehensive taxonomy. arXiv preprint arXiv:2005.03675, p. 1 (2020)
5. Dehghan, A., Kamiński, B., Prałat, P.: Node structural representation learning using local signature matrix embedding [LSME] (2022, work in progress)
6. Dehghan-Kooshkghazi, A., Kamiński, B., Kraiński, Ł., Prałat, P., Théberge, F.: Evaluating node embeddings of complex networks. J. Complex Netw. **10**(4), cnac030 (2022)
7. Donnat, C., Zitnik, M., Hallac, D., Leskovec, J.: Learning structural node embeddings via diffusion wavelets. In: Proceedings of the 24th ACM SIGKDD International Conference on Knowledge Discovery & Data Mining, pp. 1320–1329 (2018)
8. Everett, M.G., Borgatti, S.P.: Unpacking Burt's constraint measure. Soc. Netw. **62**, 50–57 (2020)
9. Grover, A., Leskovec, J.: node2vec: scalable feature learning for networks. In: Proceedings of the 22nd ACM SIGKDD International Conference on Knowledge Discovery and Data Mining, pp. 855–864 (2016)
10. Hamilton, W.L., Ying, R., Leskovec, J.: Representation learning on graphs: methods and applications. arXiv preprint arXiv:1709.05584 (2017)
11. Hasan, M.A., Zaki, M.J.: A survey of link prediction in social networks. In: Aggarwal, C. (ed.) Social Network Data Analytics, pp. 243–275. Springer, Boston (2011). https://doi.org/10.1007/978-1-4419-8462-3_9
12. Henderson, K., et al.: RolX: structural role extraction & mining in large graphs. In: Proceedings of the 18th ACM SIGKDD International Conference on Knowledge Discovery and Data Mining, pp. 1231–1239 (2012)
13. Kamiński, B., Kraiński, Ł, Prałat, P., Théberge, F.: A multi-purposed unsupervised framework for comparing embeddings of undirected and directed graphs. Netw. Sci. **10**, 323–346 (2022)

14. Kamiński, B., Prałat, P., Théberge, F.: A scalable unsupervised framework for comparing graph embeddings. In: Kamiński, B., Prałat, P., Szufel, P. (eds.) WAW 2020. LNCS, vol. 12091, pp. 52–67. Springer, Cham (2020). https://doi.org/10.1007/978-3-030-48478-1_4
15. Kamiński, B., Prałat, P., Théberge, F.: An unsupervised framework for comparing graph embeddings. J. Complex Netw. 8(5), cnz043 (2020)
16. Kamiński, B., Prałat, P., Théberge, F.: Mining Complex Networks. Chapman and Hall/CRC, London (2021)
17. Mikolov, T., Chen, K., Corrado, G., Dean, J.: Efficient estimation of word representations in vector space. arXiv preprint arXiv:1301.3781 (2013)
18. Neville, J., Jensen, D.: Iterative classification in relational data. In: Proceedings of the AAAI-2000 Workshop on Learning Statistical Models from Relational Data, pp. 13–20 (2000)
19. Pankratz, B., Kamiński, B., Prałat, P.: Community detection supported by node embeddings. In: Cherifi, H., Mantegna, R.N., Rocha, L.M., Cherifi, C., Micciche, S. (eds.) Complex Networks and Their Applications XI. Studies in Computational Intelligence, vol. 1078, pp. 221–232. Springer, Cham (2022). https://doi.org/10.1007/978-3-031-21131-7_17
20. Perozzi, B., Al-Rfou, R., Skiena, S.: DeepWalk: online learning of social representations. In: Proceedings of the 20th ACM SIGKDD International Conference on Knowledge Discovery and Data Mining, pp. 701–710 (2014)
21. Ribeiro, L.F., Saverese, P.H., Figueiredo, D.R.: struc2vec: learning node representations from structural identity. In: Proceedings of the 23rd ACM SIGKDD International Conference on Knowledge Discovery and Data Mining, pp. 385–394 (2017)
22. Stolman, A., Levy, C., Seshadhri, C., Sharma, A.: Classic graph structural features outperform factorization-based graph embedding methods on community labeling. In: Proceedings of the 2022 SIAM International Conference on Data Mining (SDM), pp. 388–396. SIAM (2022)

Modularity Based Community Detection in Hypergraphs

Bogumił Kamiński[1], Paweł Misiorek[2], Paweł Prałat[3(✉)],
and François Théberge[4]

[1] Decision Analysis and Support Unit, SGH Warsaw School of Economics,
Warsaw, Poland
`bogumil.kaminski@sgh.waw.pl`
[2] Institute of Computer Sciences, Poznan University of Technology, Poznan, Poland
`pawel.misiorek@put.poznan.pl`
[3] Department of Mathematics, Toronto Metropolitan University,
Toronto, ON, Canada
`pralat@torontomu.ca`
[4] Tutte Institute for Mathematics and Computing, Ottawa, ON, Canada
`theberge@ieee.org`

Abstract. In this paper, we make a significant step toward designing a scalable community detection algorithm using hypergraph modularity function. The main obstacle with adjusting the initial stage of the classical **Louvain** algorithm is dealt via carefully adjusted linear combination of the graph modularity function of the corresponding two-section graph and the desired hypergraph modularity function. It remains to properly tune the algorithm and design a mechanism to adjust the weights in the modularity function (in an unsupervised way), depending on how often nodes in one community share hyperedges with nodes from other communities. It will be done in the journal version of this paper.

Keywords: Community Detection Algorithm · Hypergraphs · Modularity Function

1 Introduction

Many networks that are currently modelled as graphs would be more accurately modelled as hypergraphs. This includes the collaboration network in which nodes correspond to researchers and hyperedges correspond to papers that consist of nodes associated with researchers that co-author a given paper.

After many years of intense research using graph theory in modelling and mining complex networks [13,15,21,31], hypergraphs start gaining considerable traction [2–4,6]. Standard but important questions in network science are revisited in the context of hypergraphs. However, hypergraphs also create brand new questions which did not have their counterparts for graphs. For example, how hyperedges overlap in empirical hypergraphs [30]? Or how the existing patters in a hypergraph affect the formation of new hyperedges [16]?

M. Dewar et al. (Eds.): WAW 2023, LNCS 13894, pp. 52–67, 2023.
https://doi.org/10.1007/978-3-031-32296-9_4

In this paper we concentrate on the classical problem of *community detection* in networks that can be represented using hypergraphs [1,5,8,9,18,19,25,26,34, 35]. Despite the fact that currently there is a vivid discussion around hypergraphs, the theory and tools are still not sufficiently developed to tackle this problem directly within this context. Indeed, researchers and practitioners often create the 2-section graph of a hypergraph of interest (that is, replace each hyperedge with a clique) and apply classical tools designed for graphs. After moving to the 2-section graph, one clearly loses some information about hyperedges of size greater than two and so there is a common belief that one can do better by using the knowledge of the original hypergraph.

As mentioned earlier, there are some recent attempts to deal with hypergraphs in the context of clustering. For example, Kumar et al. [25,26] still reduce the problem to graphs but use original hypergraphs to iteratively adjust weights to encourage some hyperedges to be included in some cluster but discourage other ones (this process can be viewed as separating signal from noise). Moreover, in [18,19] a number of extensions of the classic null model for graphs are proposed that can potentially be used by true hypergraph algorithms.

Unfortunately, there are many ways such extensions can be done depending on how often nodes in one community share hyperedges with nodes from other communities. We believe that the underlying process that governs *pureness* of community hyperedge is something that varies between networks at hand and also potentially depends on the hyperedge sizes. Let us come back to the collaboration network we discussed earlier. Hyperedges associated with papers written by mathematicians might be more homogeneous and smaller in comparison with those written by medical doctors who tend to work in large and multidisciplinary teams. Moreover, in general, papers with a large number of co-authors tend to be less homogeneous, and other patterns can be identified [16]. In this paper, we assume that the user has a knowledge which of the null models should be chosen to analyze a given hypergraph at hand and, as a result, wants to use the appropriate modularity function to identify communities in a hypergraph. Eventually, our clustering algorithm will be able to automatically decide which extension should be used but details will be provided in the journal version of this paper.

A significant challenge in optimizing modularity functions is that these objective functions have their domains defined over all partitions of the set of nodes and they are known to be extremely difficult to optimize. One of the most popular and efficient heuristic methods for modularity optimization for graphs is the **Louvain** algorithm. In this paper, we show how this algorithm can be adapted to optimize hypergraph modularity. One of the main challenges is the fact that, when hyperedges of size two (edges) or three are not present in the hypergraph, then the **Louvain** algorithm immediately gets stuck in its local minimum. Moreover, even if there are a few hyperedges of size two or three, then the algorithm may get stuck almost immediately. Hence, in such situations, one cannot simply start optimizing the hypergraph modularity right from the beginning. More importantly, we observe that even if hyperedges of size two are present in the hypergraph, the algorithm often converges to a local optimum that is of low quality. In order to address these two problems, we propose a method that works reasonably well in practice in which we optimize a weighted average of the 2-section graph modularity function and the hypergraph modularity function. For that we adjust the **Louvain** algorithm in such a way that the weight of the hypergraph modularity function increases during the optimization process.

The paper is structured as follows. We first introduce the necessary notation; in particular, we state the definitions of graph and hypergraph modularity functions (Sect. 2). Next, we discuss the classical **Louvain** algorithm and explain why it is difficult to adjust it to directly optimize hypergraph modularity. Following this, we describe our algorithm that is considering a linear combination of the 2-section graph modularity and the hypergraph modularity as objective function, and explain its implementation challenges (Sect. 3). Then, we present the results of numerical experiments of using the proposed algorithm on synthetic hypergraphs (Sect. 4). The paper is concluded with a summary of outlooks for further research in this area that will be addressed in the journal version of this paper (Sect. 5).

2   Modularity Functions

Let us start with some basic definitions. In the hypergraph $H = (V, E)$, each hyperedge $e \in E$ is a multiset of V of any cardinality $d \in \mathbb{N}$. Multisets in the context of hypergraphs are natural generalization of loops in the context of graphs. Hypergraphs are natural generalization of graphs in which edge is a multiset of size two. Even though H does not always contain multisets, it is convenient to allow them as they may appear in the random hypergraph that will be used as the null model to "benchmark" the edge contribution component of the modularity function. It will be convenient to partition the hyperedge set E into $\{E_1, E_2, \ldots\}$, where E_d consists of hyperedges of size d. As a result, hypergraph H can be expressed as the disjoint union of d-*uniform hypergraphs* $H = \bigcup H_d$, where $H_d = (V, E_d)$. As for graphs, $\deg_H(v)$ is the degree of node v, that is, the number of hyperedges v is a part of (taking into account the fact that hyperedges are multisets). Finally, the volume of a subset of nodes $A \subseteq V$ is $\mathrm{vol}_H(A) = \sum_{v \in A} \deg_H(v)$.

Graph Modularity. The definition of modularity for graphs was first introduced by Newman and Girvan in [33]. Despite some known issues with this function such as the "resolution limit" reported in [14], many popular algorithms for partitioning nodes of large graphs use it [11,28,32] and perform very well. The modularity function favours partitions of the set of nodes of a graph G in which a large proportion of the edges fall entirely within the parts (often called clusters), but benchmarks it against the expected number of edges one would see in those parts in the corresponding Chung-Lu random graph model [10] which generates graphs with the expected degree sequence following exactly the degree sequence in G.

Formally, for a graph $G = (V, E)$ and a given partition $\mathbf{A} = \{A_1, A_2, \ldots, A_k\}$ of V, the *modularity function* is defined as follows:

$$q_G(\mathbf{A}) = \sum_{A_i \in \mathbf{A}} \frac{e_G(A_i)}{|E|} - \sum_{A_i \in \mathbf{A}} \left(\frac{\mathrm{vol}_G(A_i)}{\mathrm{vol}_G(V)} \right)^2, \tag{1}$$

where $e_G(A_i)$ is the number of edges in the subgraph of G *induced by* set A_i. The first term in (1), $\sum_{A_i \in \mathbf{A}} e_G(A_i)/|E|$, is called the *edge contribution* and it computes the fraction of edges that fall within one of the parts. The second one, $\sum_{A_i \in \mathbf{A}} (\mathrm{vol}_G(A_i)/\mathrm{vol}_G(V))^2$, is called the *degree tax* and it computes the expected fraction of edges that do the same in the corresponding random graph (the null model). The modularity measures the deviation between the two.

The maximum *modularity* $q^*(G)$ is defined as the maximum of $q_G(\mathbf{A})$ over all possible partitions $\mathbf{A}$ of V; that is, $q^*(G) = \max_{\mathbf{A}} q_G(\mathbf{A})$. In order to maximize $q_G(\mathbf{A})$ one wants to find a partition with large edge contribution subject to small degree tax. If $q^*(G)$ approaches 1 (which is the trivial upper bound), we observe a strong community structure; conversely, if $q^*(G)$ is close to zero (which is the trivial lower bound), there is no community structure. The definition in (1) can be generalized to weighted edges by replacing edge counts with sums of edge weights.

Using Graph Modularity for Hypergraphs. Given a hypergraph $H = (V, E)$, it is common to transform its hyperedges into complete graphs (cliques), the process known as forming the 2-section of H, graph $H_{[2]}$, on the same set of nodes as H. For each hyperedge $e \in E$ with $|e| \geq 2$ and weight $w(e)$, $\binom{|e|}{2}$ edges are formed, each of them with weight of $w(e)/(|e| - 1)$. While there are other natural choices for the weights (such as the original weighting scheme $w(e)/\binom{|e|}{2}$ that preserves the total weight), this choice ensures that the degree distribution of the created graph matches the one of the original hypergraph H [25,26]. Moreover, let us also mention that it also nicely translates a natural random walk on H into a random walk on the corresponding $H_{[2]}$. As hyperedges in H usually overlap, this process creates a multigraph. In order for $H_{[2]}$ to be a simple graph, if the same pair of vertices appear in multiple hyperedges, the edge weights are added together.

One of the approaches for finding communities in hypergraphs that practitioners use is to apply **Louvain** algorithm to graph $H_{[2]}$. Despite the fact that this procedure is simple, it has a drawback that the 2-section graph looses some potentially useful information. Therefore, it is desired to define modularity for a hypergraph and aim to optimize it directly.

Hypergraph Modularity. For edges of size greater than 2, several definitions can be used to quantify the edge contribution for a given partition $\mathbf{A}$ of the set of nodes. As a result, the choice of hypergraph modularity function is not unique. It depends on how strongly one believes that a hyperedge is an indicator that some of its vertices fall into one community. The fraction of nodes of a given hyperedge that belong to one community is called its *homogeneity* (provided it is more than 50%). In one extreme case, all vertices of a hyperedge have to belong to one of the parts in order to contribute to the modularity function; this is the *strict* variant assuming that only homogeneous hyperedges provide information about underlying community structure. In the other natural extreme variant, the *majority* one, one assumes that edges are not necessarily homogeneous and so a hyperedge contributes to one of the parts if more than 50% of its vertices belong to it; in this case being over 50% is the only information that is considered relevant for community detection. All variants in between guarantee that hyperedges contribute to at most one part. Once the variant is fixed, one needs to benchmark the corresponding edge contribution using the degree tax computed for the generalization of the Chung-Lu model to hypergraphs proposed in [18].

The hypergraph modularity function is controlled by *hyper-parameters* $w_{c,d} \in [0,1]$ ($d \geq 2$, $\lfloor d/2 \rfloor + 1 \leq c \leq d$). For a fixed set of hyper-parameters, we define

$$q_H(\mathbf{A}) = \sum_{d \geq 2} \sum_{c=\lfloor d/2 \rfloor + 1}^{d} w_{c,d} \; q_H^{c,d}(\mathbf{A}), \tag{2}$$

where

$$q_H^{c,d}(\mathbf{A}) = \frac{1}{|E|} \sum_{A_i \in \mathbf{A}} \left(e_H^{d,c}(A_i) - |E_d| \cdot \Pr\left(\mathrm{Bin}\left(d, \frac{\mathrm{vol}(A_i)}{\mathrm{vol}(V)} \right) = c \right) \right);$$

$e_H^{d,c}(A_i)$ is the number of hyperedges of size d that have exactly c members in A_i, and $\mathrm{Bin}(d,p)$ is the binomial random variable.

Hyper-parameters $w_{c,d}$ give us a lot of flexibility and allow to value some edges more than others depending on their size and homogeneity. However, there are three natural hyper-parameters that one might consider, yielding three modularity functions to optimize:

- *strict modularity*: $w_{d,d} = 1$ and $w_{c,d} = 0$ for $\lfloor d/2 \rfloor + 1 \leq c < d$,
- *linear modularity*: $w_{c,d} = c/d$ for $\lfloor d/2 \rfloor + 1 \leq c \leq d$,
- *majority modularity*: $w_{c,d} = 1$ for $\lfloor d/2 \rfloor + 1 \leq c \leq d$.

3   Hypergraph Modularity Optimization Algorithm

3.1   Louvain Algorithm

Let us start by introducing one of the mostly used unsupervised algorithms for detecting communities in graphs, namely, the **Louvain** algorithm [7]. It is a hierarchical clustering algorithm that tries to optimize the modularity function we described in Sect. 2.

In this algorithm, small communities are first found by optimizing modularity locally on all nodes. Then, each small community is grouped into one node and the original step is repeated on a smaller graph. The process stops when no improvement on the modularity function can be further achieved.

One pass of the algorithm consists of two phases that are repeated iteratively. Initially, each node in the network is assigned to its own community. For
each node v, we consider all neighbours u of v and compute the change in the
modularity function if v is removed from its own community and moved into the
community of u. It is important to mention that this value can be easily and efficiently calculated without the need to recompute the modularity function from
scratch. Once all the communities that v could belong to are considered, v is
placed into the community that resulted in the largest increase of the modularity function. If no increase is possible, v remains in its original community. The
process is repeated for the remaining nodes following a given (typically random)
permutation of nodes. If no increase is possible after considering all nodes, a
local maximum value is achieved and the first phase ends.

During the second phase, the algorithm contracts all nodes that belong to one
community into a single node. All edges within that community are replaced by
a single weighted loop. Similarly, all edges between two communities are replaced
by a single weighted edge. Once the new network is created, the second phase
ends. The resulting graph is typically much smaller than the original graph. As a
result, the first pass is typically the most time consuming part of the algorithm.

3.2   Challenges with Adjusting the Algorithm to Hypergraphs

One could try to directly apply the **Louvain** algorithm to optimize hypergraph
modularity, since in both cases the goal is to find a partition of the set nodes.
However, as the algorithm moves only one node at a time, it creates a problem
in the case of hypergraphs.

Consider, for example, a hypergraph in which all hyperedges have size at least
four. In this case, regardless which two nodes u and v are considered for possible
merging into one community, the edge contribution would not change (that is,
it would stay equal to zero), even if u and v are part of some hyperedge. (Recall
that only hyperedges with majority of nodes from the same community may
affect the edge contribution). On the other hand, the degree tax would increase
after such a move and, as a result, the modularity function would decrease.
Therefore, no move would be made and the algorithm would get immediately
stuck. This problem can be referred to as a *lift off from the ground* problem.

The above, extreme, situation is not the only problem one should be aware
of. Consider this time a hypergraph that consists of a mixture of hyperedges
of various sizes, including edges of size two. In this scenario there is no problem with lifting off from the ground but small hyperedges clearly play a much
more important role than large ones during the initial merging in the first phase
of the algorithm. On the other hand, very large hyperedges would be mostly
ignored. This behaviour is not desirable either. In order to illustrate a potential danger, consider a hypergraph representing interactions between researchers
at some institution. Nodes in this hypergraph correspond to researchers and
hyperedges correspond to meetings of some groups of people. For simplicity,
assume that there are two communities, say, faculty of science and faculty of

engineering. Many hyperedges within the two communities are large (e.g. hyperedges associated with departmental meetings) whereas hyperedges between the two communities are mostly of size two (e.g. two members of different teams meet individually from time to time). In this scenario, the algorithm would start merging people from different communities during the first phase.

Finally, let us note that one could alternatively consider modifying the algorithm and allow for not only merging two nodes into one community in a single move but entire hyperedges. Again, this does not seem to be desirable as hyperedges might consist of members from different communities and so such operations would generate many incorrect merges too fast.

3.3   Our Approach to Hypergraph Modularity Optimization: h-Louvain

In order to overcome the above mentioned challenges, we want to design an algorithm that, as in the classical **Louvain** algorithm, merges single pairs of nodes while, at the same time, takes into account information stored in hyperedges of all sizes. To that end we propose to optimize a linear combination of the hypergraph modularity $q_H(\mathbf{A})$ and the graph modularity of the corresponding 2-section graph $H_{[2]}$, that is, optimize function

$$q(\mathbf{A}, \alpha) := \alpha \cdot q_H(\mathbf{A}) + (1 - \alpha) \cdot q_{H_{[2]}}(\mathbf{A}), \qquad (3)$$

where $\alpha \in [0, 1]$. For simplicity, we will refer to our algorithm as **h-Louvain**.

To understand the motivation behind this approach, let us observe the following. The hypergraph modularity, Eq. (2), is flexible and may approximate well the graph modularity for the corresponding 2-section graph $H_{[2]}$. Indeed, if c vertices of a hyperedge e of size d and weight $w(e)$ fall into one part of the partition $\mathbf{A}$, then the contribution to the graph modularity is $w(e)\binom{c}{2}/(|e| - 1)$ (in the variant of the 2-section where the degrees are preserved) or $w(e)\binom{c}{2}/\binom{|e|}{2} \approx w(e)(c/|e|)^2$ (if the total weight is preserved). Hence, the hyper-parameters of the hypergraph modularity can be adjusted to approximate $H_{[2]}$ modularity. The only difference is that (2) does not allow to include contributions from parts that contain at most $d/2$ vertices which still contributes to the graph modularity of $H_{[2]}$.

The observation justifies using $q(\mathbf{A}, \alpha)$ for optimizing the hypergraph modularity. It is a linear combination of the actual hypergraph modularity we want to optimize, $q_H(\mathbf{A})$, and an approximation of the hypergraph modularity for special values of hyper-parameters and without the restriction of hyperedge contribution, $q_{H_{[2]}}(\mathbf{A})$. The benefit of the second part is that it is sensitive to merging two nodes and so it always gives some indication of how nodes should be merged. In short, it resolves the *lifting off from the ground* problem. If α is close to zero, then we concentrate mostly on the approximation part, while if α is close to one, then we mostly concentrate on the actual hypergraph modularity we aim to optimize.

The above discussion leads us to the conclusion that the parameter $\alpha \in [0, 1]$ should be appropriately tuned during the algorithm. The main questions are: a) when the change should be made, and b) what values of this parameter should be used? The main goal of this paper is to answer these two questions.

Given the theoretical derivation we presented above, the following hypotheses (that will be verified in Sect. 4) can be formulated:

- the optimization process should be started with low values of the parameter α (to let the process *lift off from the ground*) and then it should be gradually increased till it reaches one by the end of the process (since for this value, we reduce the problem to optimizing the function we actually want to optimize);
- the algorithm should start increasing parameter α when the communities induce enough edges so that merging additional nodes makes a difference in the edge contribution of the q_H function value; this, in particular, means that since the strict hypergraph modularity pays attention to only pure hyperedges (all members belong to one community), in this case, the algorithm needs to start with lower values of α and increase it slower than for the majority or the linear counterparts of the hypergraph modularity for which it is enough that over 50% of nodes in some hyperedge are captured in one community.

4   Results

4.1   Synthetic Hypergraph Model: h-ABCD

There are very few datasets with ground-truth identified and labelled. As a result, there is need for synthetic random graph models with community structure that resemble real-world networks in order to benchmark and tune clustering algorithms that are unsupervised by nature. Since we are at the initial stage of developing our algorithm, we concentrate on experiments on synthetic networks but real-world ones will be investigated in the journal version of this paper.

The situation for graphs is rather clear. The **LFR** (Lancichinetti, Fortunato, Radicchi) model [27, 29] generates networks with communities and at the same time it allows for the heterogeneity in the distributions of both node degrees and of community sizes. It became a standard and extensively used method for generating artificial networks. The **A**rtificial **B**enchmark for **C**ommunity **D**etection (**ABCD**) [20] was recently introduced and implemented[1], including a fast implementation[2] that uses multiple threads (**ABCDe**) [24]. Undirected variant of **LFR** and **ABCD** produce graphs with comparable properties but **ABCD/ABCDe** is faster than **LFR** and can be easily tuned to allow the user to make a smooth transition between the two extremes: pure (disjoint) communities and random graph with no community structure. Moreover, it is easier to analyze theoretically—for example, in [17] various theoretical asymptotic properties of the **ABCD** model are investigated including the modularity function, arguably,

---

[1] https://github.com/bkamins/ABCDGraphGenerator.jl/.
[2] https://github.com/tolcz/ABCDeGraphGenerator.jl/.

the most important graph property of networks in the context of community detection.

Situation for hypergraphs is not as clear as for graphs. There are not only very few real-world datasets available to use but also there are not so many synthetic hypergraphs one can use. Fortunately, the building blocks in the **ABCD** model are flexible and may be adjusted to satisfy different needs. For example, the model was adjusted to include potential outliers in [23] resulting in **ABCD+o** model. Adjusting the model to hypergraphs is more complex but it was also done recently [22] resulting in **h-ABCD** model. We will use this model for our experiments.

4.2 Exhaustive Search for the Best Strategy

As discussed in Sect. 3, one hypothesis that was identified for our **h-Louvain** algorithm is that one should avoid decreasing the values of α in $q(\mathbf{A}, \alpha)$ (see Eq. (3)) as the algorithm progresses. In this first set of experiments, we consider **h-ABCD** synthetic hypergraphs on 1,000 nodes of degrees in the range $[5, 20]$ (with average around 8) and community sizes in the range $[10, 30]$ (with average around 18), where both distributions follow a power law. The hyperedges are of size between 2 and 5, inclusively, and the linear option for community hyperedges in the **h-ABCD** generator. The noise level is set at $\xi = 0.3$, the proportion of hyperedges which are sampled randomly from the set of all nodes, regardless of community memberships.

We ran an exhaustive search for every sequence of length 5 for the values of α, namely $(\alpha_1, \ldots, \alpha_5)$ which are chosen from the set $\{0, 0.25, 0.5, 0.75, 1\}^5$. For each sequence, we run a test with 5 different random seeds, for a total of $5 \cdot 5^5 = 15{,}625$ distinct runs of the **h-Louvain** algorithm.

One key question is when the value of parameter α should be changed, that is, when to move from α_i to α_{i+1} for $1 \leq i \leq 4$. After running several empirical tests, we reached the conclusion that the best results are achieved when the change is made when the number of communities reaches n/z^i, where n equals the initial number of nodes and $z \in \mathbb{R}$, $z > 1$, is a tuneable parameter. In our experiments, parameter z is set experimentally to $z = 2.3$ but in the final version of the algorithm its value will be appropriately tuned and, in particular, it will depend on the hyperedge size distribution in a given hypergraph at hand. One instance of running the **h-Louvain** algorithm is shown in Fig. 1, where we show the evolution of two quantities: the number of communities and the modularity functions. More examples can be found in the associated appendix available online[3].

---

[3] https://math.torontomu.ca/~pralat/research.html.

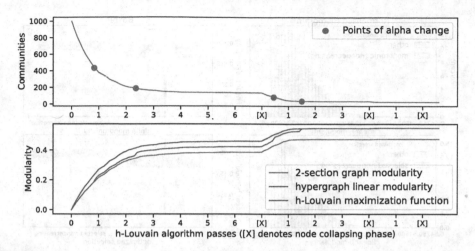

Fig. 1. Visualisation of the **h-Louvain** process for the sequence $(0.0, 0.25, 0.5, 0.75, 1)$ of α changes and maximization of the **linear** hypergraph modularity. The x-axis corresponds to the passes of the algorithm. For each pass, the iterations (based on checking all nodes in random order) of the modularity optimization phase are denoted by consecutive numbers, whereas the phase of node collapsing (community aggregation) is marked by [X].

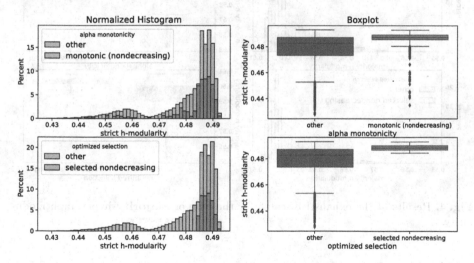

Fig. 2. Results of the exhaustive search for the case of **strict** hypergraph modularity.

In Figs. 2, 3 and 4, we show the results of the exhaustive search using, respectively, three different hypergraph modularity functions: strict, linear, and majority. In the top plots, we compare the resulting modularity values for monotonic (that is, non-decreasing) sequences of α's versus all remaining sequences, while in the bottom plots, we restrict the non-decreasing sequences to the ones where $\alpha_3 < 1$ and $\alpha_5 = 1$, thus forcing sequences of non-decreasing values spanning

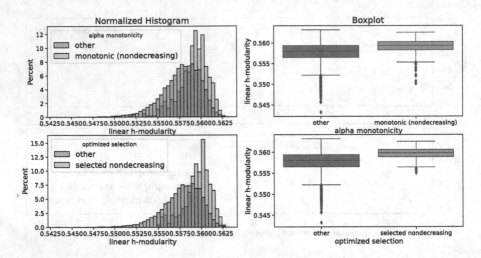

Fig. 3. Results of the exhaustive search for the case of **linear** hypergraph modularity.

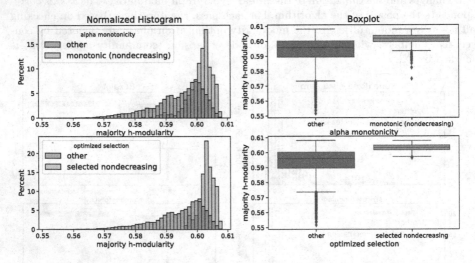

Fig. 4. Results of the exhaustive search for the case of **majority** hypergraph modularity.

a wider range. For example, this avoids the non-decreasing sequence in which $\alpha_i = 1$ for all i. Based on those figures, we see that the non-decreasing sequences generally yield better results, and forcing the extra conditions greatly reduces the variability of the results. We also notice that the gain in choosing such a strategy is more visible in the case of the strict modularity, and less so for the linear modularity. This is to be expected as the linear hypergraph modularity bears more similarity to the 2-section graph modularity function.

4.3   Comparing Basic Policies for Different Modularity Functions

For the following experiments, we consider the **h-ABCD** model with the same set of parameters used in Sect. 4.2 and with three different ways to generate community edges: (i) strict, (ii) linear, and (iii) majority. In each case, we generated 10 different hypergraphs (starting with different random seeds).

We consider simple policies where we use some value α_1 for the first pass of the **h-Louvain** algorithm, α_2 for the second pass, and α_3 for all subsequent passes. Note that in all runs, there were never more than 5 passes. Thus, for example, policy $(\alpha_1, \alpha_2, \alpha_3) = (0, 0, 0)$ amounts to optimizing the 2-section graph modularity during the entire process, while policy $(1, 1, 1)$ amounts to optimizing the hypergraph modularity throughout. Each policy is tested 5 times with different random seeds for each of the 10 hypergraphs. We considered 19 different policies either with constant α_i, or non-decreasing sequences with $\alpha_3 = 1$.

We use Friedman-Nemenyi statistical test described in detail in [12] to validate whether the performance difference between compared policies is statistically significant. We follow the same procedure for the three investigated cases, i.e., strict, linear, and majority modularity optimization. First, for every generated hypergraph, we rank the policies based on the average modularity obtained in 5 runs so that the policy with the highest score is ranked 1 while the one with the worst score is ranked 19. Then, the Friedman test based on the average ranks is used to analyze if the differences between all compared policies on multiple hypergraphs are statistically significant. For each investigated case, the calculated Friedman rank sum test statistics was greater than the critical value for the assumed level of confidence $\alpha = 0.05$, so the conclusion is that the null hypothesis that there is no difference between the policies could be rejected, and the post-hoc Nemenyi test can be conducted. The Nemenyi test aims to investigate the difference between each pair of individual policies. According to the test assumptions, the difference between a given pair of policies is regarded as statistically significant if it is bigger than the critical difference CD (for an assumed confidence level 0.05) calculated based on formulas presented in [12] (in our case equal to 2.03). We used the most popular way to visualize the Nemenyi test results proposed originally in [12]. The three diagrams presented in Figs. 5, 6 and 7 illustrate the ranked performances of the compared policies along with the critical difference CD when optimizing strict, linear, and majority modularity functions. The central axis presented in the diagrams is used to plot the average ranks of compared policies sorted in decreasing order (with the best policy at the right). Moreover, in order to support the results' interpretation, the horizontal bolt lines are added below the main axis to connect the groups of policies that are not significantly different.

In general, when the strict modularity is optimized, it is important to start with low values of the parameter α, while this is not as important when optimizing linear or majority modularity functions. Other general observations are that it is best to increase α_3 to its maximum value of 1, and never to start with $\alpha_1 = 1$ right away.

Further results using the other **h-ABCD** hypergraphs as well as more numerical details can be found in the associated appendix available online[4]. The conclusions remain the same.

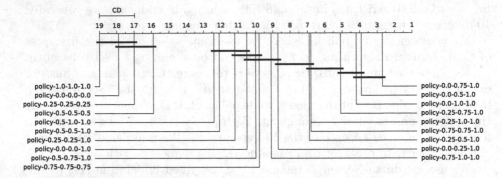

Fig. 5. Visualization of Friedman and Nemenyi test results for **strict modularity** with **linear h-ABCD**.

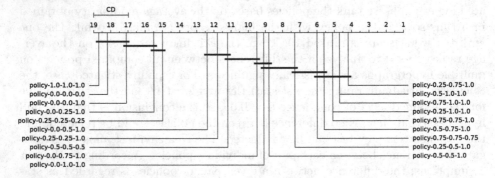

Fig. 6. Visualization of Friedman and Nemenyi test results for **linear modularity** with **linear h-ABCD**.

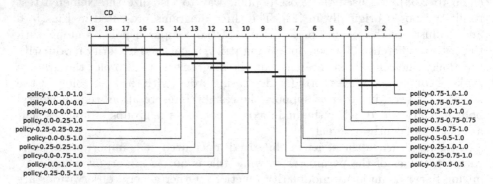

Fig. 7. Visualization of Friedman and Nemenyi test results for **majority modularity** with **linear h-ABCD**.

---

[4] https://math.torontomu.ca/~pralat/research.html.

5   Conclusions

In this paper, we proposed a modification of the classical **Louvain** algorithm that allows us to optimize the hypergraph modularity, **h-Louvain**. Our approach is to optimize a weighted average of the 2-section graph modularity and the hypergraph modularity, with an increasing weight of hypergraph modularity component as the optimization process progresses. We have presented both theoretical arguments as well as empirical evidence that the approach of increasing the weight of hypergraph modularity component improves the results of the optimization process in comparison to trying to optimize hypergraph modularity directly using the **Louvain** algorithm, or using a different weight change policy.

In this proceeding version of the paper, we concentrate on presenting the main ideas behind the proposed algorithm and the results showing that, indeed, it gives performance improvements. However, the key element of the algorithm is the schedule how the weight of the hypergraph modularity component should be changed (that is, what value should be taken and when to make the change). In the initial experiments we selected parameter α (the weight) from a fixed set of possible values and change it in discrete time steps governed by parameter z (fixed in our experiments). In the extended, journal version of this paper, more sophisticated algorithm will be presented and experimented with that will perform auto-tuning of these parameters based on various hypergraph characteristics and the type of hypergraph modularity that is optimized.

Additionally, let us mention about another important and interesting aspect. Since in **h-Louvain** the optimization process is stochastic by nature, the results of a single optimization pass can be easily improved by running many such optimizations in parallel. Therefore, an important extension to the algorithm is for allowing it to learn how to dynamically set the tuneable parameters when multiple optimization processes are executed.

One final extension that we plan to do is to allow for auto-discovery of which version of modularity function best fits the analyzed hypergraph. In the current version of the algorithm the user has to specify this information explicitly.

References

1. Ahn, K., Lee, K., Suh, C.: Hypergraph spectral clustering in the weighted stochastic block model. IEEE J. Sel. Top. Signal Process. **12**(5), 959–974 (2018)
2. Battiston, F., et al.: Networks beyond pairwise interactions: structure and dynamics. Phys. Rep. **874**, 1–92 (2020)
3. Benson, A.R., Abebe, R., Schaub, M.T., Jadbabaie, A., Kleinberg, J.: Simplicial closure and higher-order link prediction. Proc. Natl. Acad. Sci. **115**(48), E11221–E11230 (2018)
4. Benson, A.R., Gleich, D.F., Higham, D.J.: Higher-order network analysis takes off, fueled by classical ideas and new data. arXiv preprint arXiv:2103.05031 (2021)

5. Benson, A.R., Gleich, D.F., Leskovec, J.: Tensor spectral clustering for partitioning higher-order network structures. In: Proceedings of the 2015 SIAM International Conference on Data Mining, pp. 118–126. SIAM (2015)
6. Benson, A.R., Gleich, D.F., Leskovec, J.: Higher-order organization of complex networks. Science **353**(6295), 163–166 (2016)
7. Blondel, V.D., Guillaume, J.L., Lambiotte, R., Lefebvre, E.: Fast unfolding of communities in large networks. J. Stat. Mech: Theory Exp. **2008**(10), P10008 (2008)
8. Chien, I., Lin, C.Y., Wang, I.H.: Community detection in hypergraphs: optimal statistical limit and efficient algorithms. In: International Conference on Artificial Intelligence and Statistics, pp. 871–879. PMLR (2018)
9. Chodrow, P.S., Veldt, N., Benson, A.R.: Generative hypergraph clustering: from blockmodels to modularity. Sci. Adv. **7**(28), eabh1303 (2021)
10. Chung Graham, F., Lu, L.: Complex graphs and networks. No. 107, American Mathematical Society (2006)
11. Clauset, A., Newman, M.E., Moore, C.: Finding community structure in very large networks. Phys. Rev. E **70**(6), 066111 (2004)
12. Demšar, J.: Statistical comparisons of classifiers over multiple data sets. J. Mach. Learn. Res. **7**, 1–30 (2006)
13. Easley, D., Kleinberg, J.: Networks, Crowds, and Markets: Reasoning About a Highly Connected World. Cambridge University Press, Cambridge (2010)
14. Fortunato, S., Barthelemy, M.: Resolution limit in community detection. Proc. Natl. Acad. Sci. **104**(1), 36–41 (2007)
15. Jackson, M.O.: Social and Economic Networks. Princeton University Press, Princeton (2010)
16. Juul, J.L., Benson, A.R., Kleinberg, J.: Hypergraph patterns and collaboration structure. arXiv preprint arXiv:2210.02163 (2022)
17. Kamiński, B., Pankratz, B., Prałat, P., Théberge, F.: Modularity of the abcd random graph model with community structure. preprint arXiv:2203.01480 (2022)
18. Kamiński, B., Poulin, V., Prałat, P., Szufel, P., Théberge, F.: Clustering via hypergraph modularity. PLoS ONE **14**(11), e0224307 (2019)
19. Kamiński, B., Prałat, P., Théberge, F.: Community detection algorithm using hypergraph modularity. In: Benito, R.M., Cherifi, C., Cherifi, H., Moro, E., Rocha, L.M., Sales-Pardo, M. (eds.) Complex Networks and Their Applications, pp. 152–163. Springer, Cham (2020). https://doi.org/10.1007/978-3-030-65347-7_13
20. Kamiński, B., Prałat, P., Théberge, F.: Artificial benchmark for community detection (ABCD)-fast random graph model with community structure. Netw. Sci. 1–26 (2021)
21. Kamiński, B., Prałat, P., Théberge, F.: Mining Complex Networks. Chapman and Hall/CRC, Boca Raton (2021)
22. Kamiński, B., Prałat, P., Théberge, F.: Hypergraph artificial benchmark for community detection (h-ABCD). arXiv preprint arXiv:2210.15009 (2022)
23. Kamiński, B., Prałat, P., Théberge, F.: Outliers in the ABCD random graph model with community structure (ABCD+O). In: Proceedings of the 11th International Conference on Complex Networks and their Applications (2022, in press)
24. Kamiński, B., Olczak, T., Pankratz, B., Prałat, P., Théberge, F.: Properties and performance of the ABCDE random graph model with community structure. Big Data Res. **30**, 100348 (2022)
25. Kumar, T., Vaidyanathan, S., Ananthapadmanabhan, H., Parthasarathy, S., Ravindran, B.: Hypergraph clustering by iteratively reweighted modularity maximization. Appl. Netw. Sci. **5**(1), 1–22 (2020)

26. Kumar, T., Vaidyanathan, S., Ananthapadmanabhan, H., Parthasarathy, S., Ravindran, B.: A new measure of modularity in hypergraphs: theoretical insights and implications for effective clustering. In: Cherifi, H., Gaito, S., Mendes, J.F., Moro, E., Rocha, L.M. (eds.) COMPLEX NETWORKS 2019. SCI, vol. 881, pp. 286–297. Springer, Cham (2020). https://doi.org/10.1007/978-3-030-36687-2_24
27. Lancichinetti, A., Fortunato, S.: Benchmarks for testing community detection algorithms on directed and weighted graphs with overlapping communities. Phys. Rev. E **80**(1), 016118 (2009)
28. Lancichinetti, A., Fortunato, S.: Limits of modularity maximization in community detection. Phys. Rev. E **84**(6), 066122 (2011)
29. Lancichinetti, A., Fortunato, S., Radicchi, F.: Benchmark graphs for testing community detection algorithms. Phys. Rev. E **78**(4), 046110 (2008)
30. Lee, G., Choe, M., Shin, K.: How do hyperedges overlap in real-world hypergraphs?-patterns, measures, and generators. In: Proceedings of the Web Conference 2021, pp. 3396–3407 (2021)
31. Newman, M.: Networks. Oxford University Press, Oxford (2018)
32. Newman, M.E.: Fast algorithm for detecting community structure in networks. Phys. Rev. E **69**(6), 066133 (2004)
33. Newman, M.E., Girvan, M.: Finding and evaluating community structure in networks. Phys. Rev. E **69**(2), 026113 (2004)
34. Yin, H., Benson, A.R., Leskovec, J.: Higher-order clustering in networks. Phys. Rev. E **97**(5), 052306 (2018)
35. Yin, H., Benson, A.R., Leskovec, J., Gleich, D.F.: Local higher-order graph clustering. In: Proceedings of the 23rd ACM SIGKDD International Conference on Knowledge Discovery and Data Mining, pp. 555–564 (2017)

Establishing Herd Immunity is Hard Even in Simple Geometric Networks

Michal Dvořák[iD], Dušan Knop[iD], and Šimon Schierreich[(✉)][iD]

Department of Theoretical Computer Science, Faculty of Information Technology,
Czech Technical University in Prague, Prague, Czech Republic
{dvora125,dusan.knop,simon.schierreich}@fit.cvut.cz

Abstract. We study the following model of disease spread in a social network. In the beginning, all individuals are either *infected* or *healthy*. Next, in discrete rounds, the disease spreads in the network from infected to healthy individuals such that a healthy individual gets infected if and only if a sufficient number of its direct neighbours are already infected.

We represent the social network as a graph. Inspired by the real-world restrictions in the current epidemic, especially by social and physical distancing requirements, we restrict ourselves to networks that can be represented as geometric intersection graphs.

We show that finding a minimal vertex set of initially infected individuals to spread the disease in the whole network is computationally hard, already on unit disk graphs. Hence, to provide some algorithmic results, we focus ourselves on simpler geometric graph families, such as interval graphs and grid graphs.

Keywords: Disease spread · TARGET SET SELECTION · intersection graphs · computational complexity

1 Introduction

In this work, we study the following deterministic model of disease spread. We are given a social network represented as a simple, undirected *underlying graph* $G = (V, E)$, a threshold function $t \colon V \to \mathbb{N}$ associating each agent with her *immunity* (or *threshold*), and a *budget* $k \in \mathbb{N}$. Our goal is to select a group $S \subseteq V$, $|S| \leq k$, of initially infected agents (a *target-set*) such that all agents get infected by the following activation process:

$$S_0 = S,$$
$$S_i = \{v \mid t(v) \leq |N(v) \cap S_{i-1}|\} \cup S_{i-1}.$$

In other words, the disease spreads in discrete rounds. A healthy agent v becomes infected if the number of neighbours already infected reaches the agent's immunity value $t(v)$. We note that once an agent is infected, she remains in this state for the rest of the process.

M. Dewar et al. (Eds.): WAW 2023, LNCS 13894, pp. 68–82, 2023.
https://doi.org/10.1007/978-3-031-32296-9_5

Dreyer and Roberts [13] studied a similar model under the name IRRE-VERSIBLE k-THRESHOLD PROCESS. Unlike our setting, in their work, the immunity value is the same for all agents. Therefore, the presented model is more general and corresponds, in fact, to the TARGET SET SELECTION problem (TSS for short) where thresholds can be agent-specific.

The TARGET SET SELECTION problem was introduced by Domingos and Richardson [33] in the context of viral marketing on social networks. Kempe, Kleinberg, and Tardos [27] later refined the problem in terms of thresholds, which is the model we follow in this work, and showed that the problem is NP-hard.

The first way to tackle the problem's complexity was aimed at the threshold function. However, Chen [6] showed that TSS remains NP-hard even if all thresholds are at most two, which extends the previous result of Dreyer and Roberts [13] who showed that the problem is NP-hard even if all thresholds are bounded by a constant $c \geq 3$. NP-hardness for majority thresholds (for every $v \in V$ we have $t(v) = \lceil \deg(v)/2 \rceil$) is due to Peleg [31].

It is easy to see that the TSS problem is solvable in polynomial time if the underlying graph has diameter one, that is, it is a complete graph [30, 32]. Chen [6] showed that the problem remains polynomial-time solvable even if the underlying graph is a tree. Later, Chiang et al. [7] proposed linear-time algorithms for block-cactus graphs, chordal graphs with all thresholds at most two, and Hamming graphs with all thresholds equal to two. Bessy et al. [3] showed that the TSS problem is solvable in polynomial time on interval graphs if all thresholds are bounded by a constant. On the other hand, the problem becomes NP-hard on graphs of diameter two [30].

Restriction of the underlying graph structure was further investigated by Ben-Zwi et al. [2]. They gave an algorithm running in $n^{\mathcal{O}(\omega)}$ time for networks with n vertices and *tree-width* bounded by ω, and showed that, under reasonable theoretical assumptions, there is no algorithm running in $n^{o(\sqrt{w})}$ time. The parametrized complexity perspective, initiated by Ben-Zwi et al. [2], was later used in multiple subsequent works [8, 14, 21, 28, 30].

Finally, Cicalese et al. [9] proposed a study of the TSS problem, where the process must stabilise within a prescribed number of rounds. They give a polynomial-time algorithm for graphs of bounded *clique-width* and a linear time algorithm for trees.

Inspired by the actual restrictions in the current epidemic, especially by social and physical distancing requirements, we study the TARGET SET SELECTION problem restricted to instances where the underlying graph is a *(unit) disk graph*.

Unit disk graphs were initially used as a natural model for a topology of ad-hoc wireless communication networks [24]. For a given graph, it is NP-hard to recognise whether the graph is a unit disk graph [5, 23, 25]. On the other hand, many computationally hard problems, such as INDEPENDENT SET or COLOURING, can be efficiently approximated for this graph family [29]. CLIQUE can be solved even in polynomial time if the disk representation is given as part of the input [10].

In our case, the disk representation models two different situations. In the first situation, the disk represents the distances that individuals must keep. In the second case, the disk represents the area in which the disease is spread by an infected individual.

As the TARGET SET SELECTION problem is notoriously hard from both exact computation and approximation point of view, it is natural to ask whether any positive result can be given if we restrict TSS to instances where the underlying graph is a unit disk graph, or if we need to restrict ourselves to even simpler graph families.

Preventing Disease Spread. In what we discussed above, our goal was to spread the disease throughout the network. This goal corresponds to establishing herd immunity. However, herd immunity is not the only way to tackle the pandemic. For example, there may be a group of individuals who are very vulnerable to the disease, and the likelihood of them dying from the disease is very high. Or, in the case of a disease with very high mortality, our goal can be to minimise the number of infected individuals.

The first case presented, where we have a group of agents that must be protected, is known as GROUP IDENTIFICATION [12,26]. Despite the fact that the problem is mostly studied in the context of opinion spread, we can easily utilise it to prevent disease spread. For us, the most significant results are the works of Yang et al. [35] and Erdélyi et al. [15], who study the control of agents to manipulate an outcome. One of such controls is agent deletion, which can be translated into the vaccination or quarantining of an agent. Deciding whether there are k agents whose deletion leads to protection of a given subset of agents is solvable in polynomial time. See also [4] for more complicated settings of GROUP IDENTIFICATION.

If we want to minimise the spread of the disease, we can employ the FIRE-FIGHTER PROBLEM [17,22]. Here, we are given a graph. Then, the fire breaks out on a set of vertices, and our goal is to protect as many vertices as possible. In each round of the process, we defend a selected vertex from being burnt, and the fire spreads to all undefended neighbours. Once the vertex is on fire or is defended, it remains so. The process stops when it stabilises, that is, when there is no new burnt vertex in any round. Fomin et al. [19] showed that the FIRE-FIGHTER PROBLEM is NP-hard even on unit disk graphs, while it is solvable in polynomial time on interval graphs.

We found the setting of preventing disease spread almost solved and hence we will not assume any disease spread prevention in the rest of this paper.

Our Contribution. In this paper, we show that the TARGET SET SELECTION problem is computationally hard even on very simple geometric graph classes. In particular, we show that TSS is NP-complete on unit disk graphs even if the threshold function is bounded by a constant $c \geq 2$, is equal to majority, or is unanimous. Hence, we focus on a study of grid graphs, which is a sub-family of unit disk graphs. For grid graphs, we show that TSS is NP-complete

Table 1. Overview of our results. The first row contains individual restrictions of the threshold function, and the first column contains assumed graph families. In the table, "NP-c" stands for "NP-complete", "P" stands for polynomial-time solvable cases, and "?" indicates an open question. All results are complemented by the reference to the appropriate statement or source.

	constant	majority	unanimous	unrestricted
interval graphs	P [3]	?	P (Theorem 3)	?
grid graphs	NP-c (Theorem 5)	NP-c (Theorem 6)	P (Theorem 2)	NP-c (Theorem 5)
unit disk graphs	NP-c (Corollary 1)	NP-c (Corollary 2)	NP-c (Theorem 1)	NP-c (Theorem 1)

for constant and majority threshold, while it is polynomial-time solvable for unanimous thresholds. Note that our results for constant thresholds sets up a clear dichotomy between tractable and intractable subclasses of intersection graph families, as TSS is known to be solvable in polynomial time on interval graphs with constant thresholds [3]. Moreover, we show that our problem is solvable in polynomial-time on interval graphs even if the threshold function is unanimous. For an overview of our results, we refer the reader to Table 1.

Paper Organisation. The remainder of this paper is organised as follows. In Sect. 2, we introduce all the definitions and notation used throughout the paper. In Sect. 3, we show the hardness and algorithmic results for TARGET SET SELECTION restricted to unanimous thresholds. Section 4 is dedicated to the variant where the maximum threshold value is bounded by a constant. At first, we give a somewhat straightforward NP-hardness proof for thresholds at most 3. Later, we give a more involved proof showing hardness even in a special case with thresholds of at most 2. In Sect. 5, we study a variant of the TARGET SET SELECTION problem with majority thresholds, and we conclude the paper with some open problems and future research directions in Sect. 6.

2 Preliminaries

A simple undirected graph is a pair $G = (V, E)$, where V is a set of vertices, and $E \subseteq \binom{V}{2}$ is a set of edges. Let u and v be two distinct vertices. If $\{u, v\} \in E$, then we call u a *neighbour* of v and vice versa. We denote the set of all neighbours of a vertex v by $N(v)$ and $|N(v)| = \deg(v)$ is the *degree* of the vertex v. A vertex of degree 1 is called a *pendant vertex*. We denote the set of all pendant vertices in a graph G by L_G. Let $c \in \mathbb{N}$ be a constant. We say that a graph G is *c-regular* if every vertex $v \in V$ has degree exactly c. A graph is *regular*, if it is c-regular for some constant $c \in \mathbb{N}$.

Definition 1 (Unit disk graph). *A graph $G = (V, E)$ is a disk graph if there exists a collection $\mathcal{D} = (D_1, \ldots, D_n)$ of $n = |V|$ closed disks in plane such that $\forall i \in [n]$ v_i is a centre of D_i, $\{v_i, v_j\} \in E$ if and only if $D_i \cap D_j \neq \emptyset$. If all disks $D_i \in \mathcal{D}$ have same radii, we call the graph a unit disk graph.*

Let G and H be two graphs. The *Cartesian product* of the graphs G and H is a graph $G \square H$ such that $V(G \square H) = V(G) \times V(H)$ and $\{(u, u'), (v, v')\}$ is an edge if and only if either $u = v$ and u' is a neighbour of v' in H, or $u' = v'$ and u is a neighbour of v in G.

Definition 2 (Grid graph). *An $n \times m$ grid is the Cartesian product of the path graphs P_n and P_m. A graph G is a grid graph if and only if it is an induced subgraph of a grid.*

Let $G = (V, E)$ be a graph. We call a set $C \subseteq V$ a *vertex cover* of the graph G, if at least one end of each edge is a member of C. In the VERTEX COVER (VC for short) problem, we are given a graph G and an integer $k \in \mathbb{N}$, and our goal is to decide whether there is a vertex cover C of size at most k. A set $I \subseteq V$ is called an *independent set*, if for every pair of distinct vertices $u, v \in I$ there is no edge connecting u and v. In the INDEPENDENT SET problem (IS for short), we are given a graph G and an integer $k \in \mathbb{N}$, and our goal is to decide whether there is an independent set I of size at least k.

Almost all our hardness reductions come from some variant of the 3-SAT problem. In this problem, we are given a propositional formula φ in conjunctive normal form (CNF) on the set $X = \{x_1, \dots, x_n\}$ of *variables*. The set of clauses is denoted $\mathcal{C} = \{C_1, \dots, C_m\}$ and each clause contains at most 3 literals. Our goal is to decide whether there is a truth assignment $\pi \colon X \to \{0, 1\}$ that satisfies φ. We say that φ is *restricted* if every variable is part of at most 3 clauses and occurs negated in exactly one clause. An *incidence graph* G_φ for the formula φ is a bipartite graph $(X \cup \mathcal{C}, E)$ with an edge between $x_i \in X$ and $C_j \in \mathcal{C}$ if and only if either $x_i \in C_j$ or $\neg x_i \in C_j$. If G_φ is planar, we say that the 3-SAT instance is *planar*. Let $C_j \in \mathcal{C}$ be a clause. If C_j contains only positive literals, we call this clause *positive*. We call C_j *negative* if it contains only negative literals. An instance of the 3-SAT problem is called *monotone* if it contains only positive and negative clauses.

3   Unanimous Thresholds

In this section, we study the special case of the TARGET SET SELECTION problem where all thresholds are equal to the degree of a vertex, that is, for every $v \in V$ we have $t(v) = \deg(v)$. Such thresholds are called *unanimous*. We strongly rely on the following easy-to-see and well-known equivalence between the TARGET SET SELECTION problem with unanimous thresholds and the VERTEX COVER problem. For the sake of completeness, we provide our own proof of the lemma.

Lemma 1. *The* TARGET SET SELECTION *problem with unanimous thresholds is equivalent to the* VERTEX COVER *problem.*

Proof. Let $\mathcal{I} = (G, k)$ be an instance of the VERTEX COVER problem. We construct an equivalent instance $\mathcal{I}' = (G', k', t)$ of the TARGET SET SELECTION problem as follows. We set $G' = G$, $k' = k$, and for every $v \in V(G')$ we set the threshold value to be equal to the degree of the vertex v, that is, $t(v) = \deg_G(v)$.

Let $\mathcal{I}$ be a *yes*-instance, and $C \subseteq V(G)$ be a vertex cover of size at most k. We claim that a set $S = C$ is a solution of $\mathcal{I}'$. It is easy to see that $|S| \leq k'$. Furthermore, in the first round of activation, all the vertices in $V(G') \setminus S$ become infected as all their neighbours are already infected. Therefore, $\mathcal{I}'$ is indeed a *yes*-instance.

In the opposite direction, let $\mathcal{I}'$ be a *yes*-instance and $S \subseteq V(G')$ be a target set of size at most k'. For every vertex $v \in V(G') \setminus S$ we have $N_{G'}(v) \cup S = N_{G'}(v)$. If not, then due to the unanimous thresholds, neither v nor $u \in N_{G'}(v) \setminus S$ become infected by the activation process. Thus, $V(G') \setminus S$ is an independent set, and S is a vertex cover of G of size at most k. $\qquad\Box$

It is known that the VERTEX COVER problem is NP-complete even on unit disk graphs [10]. By Lemma 1 we have that the VERTEX COVER problem is equivalent to the TSS problem with unanimous thresholds. Therefore, the following theorem holds.

Theorem 1. TARGET SET SELECTION *is NP-complete even if the underlying graph is a unit disk graph, the geometric representations is given, and all thresholds are unanimous.*

It follows from Theorem 1 that the general case of TSS on unit disk graphs with unrestricted threshold function is NP-complete as well.

As the TARGET SET SELECTION problem is, under reasonable complexity assumptions, computationally intractable on unit disk graphs, we focus ourselves on a subfamily of unit disks called grid graphs. In particular, we show that, in this subfamily, the TSS problem becomes tractable.

Theorem 2. TARGET SET SELECTION *can be solved in polynomial time if the underlying graph is a grid graph and all thresholds are unanimous.*

Proof. It is known that the VERTEX COVER problem is solvable in polynomial time on grid graphs [10]. Together with Lemma 1 the theorem follows. $\qquad\Box$

Finally, Bessy et al. [3] showed that TSS can be solved in polynomial time on interval graphs if all thresholds are bounded by a constant. We complement this result by showing that a set of initially infected agents can be found in linear time if all thresholds are unanimous.

Theorem 3. TARGET SET SELECTION *can be solved in linear time if the underlying graph is an interval graph and all thresholds are unanimous.*

Proof. We proceed in the same way as in the proof of Theorem 1. By Lemma 1 we find that TSS is equivalent to the VERTEX COVER problem which can be solved in linear time on interval graphs [16]. $\qquad\Box$

4   Constant Thresholds

The TARGET SET SELECTION problem seems to be intractable on unit disk graphs when the threshold function is unrestricted or unanimous. It is now natural to ask whether the problem remains hard even under some natural restrictions of the thresholds.

In this section, we show that TSS is NP-complete when the underlying graph is a unit disk graph and all thresholds are bounded by a constant $c \geq 2$. Note that the case where all thresholds are at most 1, is trivial. It is sufficient to choose one vertex per connected component of the input graph.

We begin with an auxiliary lemma showing hardness of the INDEPENDENT SET problem in 3-regular and 4-regular unit disk graphs which can be of independent interest.

Lemma 2. INDEPENDENT SET *is NP-complete even if the underlying graph is an* r-*regular unit disk graph, where* $r \in \{3, 4\}$.

Proof. We build a reduction from INDEPENDENT SET on r-regular planar graphs. Note that INDEPENDENT SET restricted to such instances remains NP-hard [18]. Let (G, k) be an instance of the INDEPENDENT SET problem where $G = (V, E)$ is planar r-regular graph, where $r \in \{3, 4\}$. We assume that we are given a rectilinear embedding[1] of G, which always exists, area of such embedding is at most $\mathcal{O}(n^2)$, and can be found in polynomial time due to the theorem of Valiant [34].

We replace each edge $e = \{u, v\} \in E$ with a path $ux_1x_2 \ldots x_{6q_e}v$, i.e., we subdivide each edge with $6q_e$ vertices and for $i \in \{0, \ldots, 2q_e - 1\}$ we replace every vertex x_{3i+2} with a clique K_{c-1} (see Fig. 1). Note that the number q_e depends on the edge e.

Let us denote the new graph with subdivided edges by $G' = (V', E')$. G' clearly remains r-regular and can be easily represented using unit disks; see Fig. 2. Note that choosing the diameter of the circles sufficiently small and, if needed, stretching or squeezing the circles along the edge, it is always possible to have a subdivision where there are $6q_e$ vertices on each edge (see Fig. 3).

Finally, set $k' = k + \sum_{e \in E} 3q_e$.

Claim 1. (G, k) is a *yes*-instance if and only if (G', k') is a *yes*-instance.

Proof. Assume that (G, k) is a yes-instance of INDEPENDENT SET and let S be an independent set in G of size at least k, that is, $|S| \geq k$. Let $\{u, v\} \in E$ be an arbitrary edge. Because S is an independent set, either $u \notin S$ or $v \notin S$. Without loss of generality, $u \notin S$ (otherwise swap u and v). It suffices to add all the vertices x_{2k+1} for $k \in \{0, 1, 2, \ldots, 3q_e - 1\}$ to S. It is obvious that S remains independent. For each edge, we added at least $3q_e$ vertices, so the new

---

[1] Rectilinear embedding of a planar graph with maximum degree 4 is planar embedding with vertices at integer coordinates and edges are drawn so that they are made up of line segments of the form $x = i$ or $y = j$, where $i, j \in \mathbb{Z}$ [34].

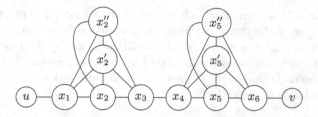

Fig. 1. An example of a construction of path subdivision in a case of 4-regular graphs.

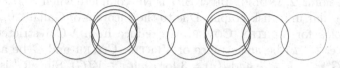

Fig. 2. An edge representation for 4-regular graph with $q = 1$. Red vertices (disks) correspond to the original vertices of an edge in G. (Color figure online)

independent set is of size of at least $k + \sum_{e \in E} 3q_e = k'$. So (G', k') is a yes-instance.

On the other hand, assume that (G', k') is a yes-instance and let S' be an independent set in G' of size at least k', that is, $|S| \geq k' = k + \sum_{e \in E} 3q_e$. Let $e = \{u, v\} \in E$ be an edge and let $ux_1x_2 \ldots x_{6q_e}v$ be its subdivision. There are two cases to consider.

- Both u, v are in S'. In this case, $|S' \cap \{x_1, \ldots, x_{6q_e}\}| \leq 3q_e - 1$ by the pigeonhole principle. We choose u or v and remove one of them from S' to make the resulting set independent in G. Also, we remove all x_i's that were in S'.
- At least one of u, v is not in S'. In this case, $|S' \cap \{x_1, \ldots, x_{6q_e}\}| \leq 3q_e$ by the pigeonhole principle, and we remove all the x_i's just as in the previous case.

In both cases, we remove at most $3q_e$ vertices for each edge. The new set contains only the vertices of G and is of size at least $k' - \sum_{e \in E(G)} 3q_e = k$ and is independent, so (G, k) is a yes instance.   ◄

It remains to say that the reduction can be done in polynomial time with respect to the input size, so the lemma follows.   □

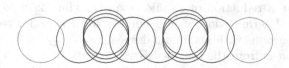

Fig. 3. Squeezing of disks.

Using Lemma 2, we can easily show that TARGET SET SELECTION remains NP-complete even if the threshold function is bounded by a constant. We show a construction for the case where all thresholds are exactly 3. It is not hard to see that our result holds for every constant $c \geq 3$; the algorithm that solves TSS with all thresholds bounded by a constant $c \geq 3$ must especially solve the case where all thresholds are exactly 3.

Theorem 4. TARGET SET SELECTION *is NP-complete even if the underlying graph is a unit disk graph and all thresholds are bounded by a constant* $c \geq 3$.

Proof. By Lemma 2, INDEPENDENT SET is NP-complete when restricted to the class of 3-regular unit disk graphs. Due to the theorem of Gallai [20], the same hardness holds for VERTEX COVER. We reduce from VC restricted to such instances. Let (G, k) be an instance of VERTEX COVER and G be a 3-regular graph. Set $G' = G, k' = k$ and $t(v) = 3$ for each $v \in V(G)$. Since G' is 3-regular this is the case of unanimous thresholds. As noted in Lemma 1, instances of TARGET SET SELECTION with unanimous thresholds are equivalent to the VERTEX COVER problem, so the theorem follows.   □

Following historical development in the study of TSS, it remains to show the complexity of the problem if all thresholds are bounded by 2 and the underlying graph is a unit disk graph.

Since we are reducing to an instance where $t(v) \leq 2$ for all vertices v, we use the following observation.

Observation 1. *Let $\mathcal{I}$ be an instance of* TARGET SET SELECTION, *and $v \in V$ be a pendant vertex with $t(v) \geq 2$. Then for any solution S it holds that $v \in S$.*

Proof. Suppose that $\mathcal{I} = (G, k, t)$ is a *yes*-instance of TSS, $S \subseteq V$ is a set of initially infected agents which is a solution of $\mathcal{I}$, and there is a vertex $v \in V \setminus S$ with $\deg(v) = 1$ and $t(v) \geq 2$. Since v has only one neighbour u, there is no way the vertex v can be infected by the activation process initiated with the set S. Hence, v must be part of S.   □

Now, we can finally provide the main result of this section showing that TSS is, in fact, hard even in a very restricted subfamily of unit disk graphs.

Theorem 5. TARGET SET SELECTION *is NP-complete even when the underlying graph is a grid graph and all thresholds are at most 2.*

Proof. We utilise a reduction of Araújo et al. [1, Thm. 2] used to proof NP-hardness of the determination of hull number for P_3- and P_3^*-convexities on planar bipartite graphs with maximum degree $\Delta \leq 4$.

The reduction is from the so-called RESTRICTED PLANAR 3-SAT problem with rectilinear representation where we are given a 3-SAT CNF formula φ with m clause $C_1, \ldots, C_m$ and n variables $x_1, \ldots, x_n$ such that each variable is part of at most 3 clauses and there is exactly one clause where x_i occurs as negated literal. This formula comes with an underlying graph G_φ, which is

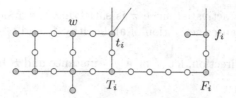

Fig. 4. Schematic representation of the variable gadget for a variable x_i. The filled vertices have threshold 2, while the empty vertices have threshold 1. Note also that the half-edges illustrate the fact that the gadget is connected with the rest of the graph only via t_i and f_i.

planar, contains two vertices ℓ_{x_i} and $\ell_{\overline{x_i}}$ for every variable x_i, one vertex c_j for every clause C_j, there is an edge between every pair ℓ_{x_i} and $\ell_{\overline{x_i}}$, and, finally, there is an edge between literal vertex ℓ_x and clause vertex c_j if and only if $x \in C_j$. The goal is to decide whether there is a truth assignment to the variables such that all clauses are satisfied. This variant of 3-SAT is known to be NP-complete [11].

At first, observe that in the graph G_φ, each vertex has a degree at most 4 and hence we can employ the result of Valiant [34] to obtain a rectilinear representation of this graph as in the proof of Lemma 2. Hence, from this point on, we assume that a rectilinear representation of G_φ is given.

The main building block of our construction is *variable gadget*. Given a variable x_i, we draw the gadget as in Fig. 4. The notable vertices of the gadget are T_i, F_i, t_i, and f_i. In the solution, the vertices T_i and F_i stands for the truth assignment of this particular variable, while the vertices t_i and f_i represent positive and negative literals, respectively, and serve to connect the variable gadgets with the respective clause gadgets. Since we start with restricted 3-SAT, there will be exactly two additional edges incident to t_i and one additional edge incident to f_i.

Our *clause gadget* is a simple path with two vertices y_j and g_j. It holds that $t(y_j) = t(g_j) = 2$. The vertex y_j is in construction connected with the corresponding literal vertices.

We construct an equivalent instance $\mathcal{J} = (G, k, t)$ of TSS as follows. For every variable x_i we replace a variable vertex $v_{x_i} \in V(G_\varphi)$ with our variable gadget X_i and each clause vertex v_{C_j} with a clause gadget Y_j. Next, we connect all literal vertices of X_i with the corresponding clause gadgets such that we add $\{t_i, y_j\}$ into E is x_i occurs as a positive literal in C_j, and $\{f_i, y_j\}$ if x_i occurs negated in C_j. To complete the construction, we use the theorem of Valiant [34] to turn G into a grid graph. Our budget is $k = 4n + n + m = 5n + m$.

Let φ be a *yes*-instance of the RESTRICTED PLANAR 3-SAT problem and π be a satisfying assignment. We create our target set S as follows. We add the vertex g_i for every clause C_i and the pendant vertices of every variable gadget X_i. Next, for every variable x_i we add either T_i if $\pi(x_i) = 1$ or F_i if $\pi(x_i) = 0$. Now $|S| = m + 4n + n = 5n$ and all the vertices become infected: every $T_i \in S$ activates the corresponding t_i and every $F_i \in S$ activates the corresponding f_i

and thus every Y_i is activated since π is satisfying assignment. It follows that activation of all Y_i causes activation of all healthy f_j and t_j, and therefore $\mathcal{J}$ is also a *yes*-instance.

In the opposite direction, let $\mathcal{J}$ be a *yes*-instance and S be a target set in $\mathcal{J}$ of size k.

Claim 2. Let $\mathcal{J}$ be a *yes*-instance and $S \subseteq V$ be a target set. For every variable gadget X_i, where $i \in [n]$, $S \cap (V(X_i) \setminus L_G) \neq \emptyset$.

Proof. Suppose that there is a variable gadget X_i such that $S \cap V(X_i) = \emptyset$ and, at some point, the vertices t_i and f_i become infected. The vertex F_i remains healthy unless the vertex T_i is healthy. Thus, we focus ourselves on the situation in the left part of the variable gadget. Once t_i is infected, it infects all healthy vertices in $N(t_i)$. In the next round, there are healthy vertices w and T_i in the second neighbourhood of t_i. But, at this point, both w and T_i have only one infected neighbour, and hence stay healthy. This contradicts that S is a target set, thus, there is always a non-empty intersection between the vertices of S and the vertices of each variable-gadget.   ◄

By Claim 2 we see that there is at least one vertex in each variable gadget that is part of S. By Observation 1, we have that g_i for every $i \in [m]$ and all pendant vertices of every variable gadget are in every S. Therefore, by the definition of k, there is exactly one non-pendant vertex vertex of each variable gadget in S.

Claim 3. Let X_i be a variable gadget and $S \subseteq V$ be a target set of size at most k with $\{T_i, F_i\} \cap S = \emptyset$. Then there is a target set S' such that $\{T_i, F_i\} \cap S' \neq \emptyset$ and $|S'| \leq |S|$.

Proof. Let u be a vertex in $S \cap (V(X_i) \setminus L_G)$. Since T_i and F_i have threshold 2, they never get infected by a vertex u alone. We recall that in every target set S there is exactly one non-pendant vertex for each variable gadget in S. Since F_i is active, both neighbours are infected. One of these is infected due to u and the second must be infected due to f_i, therefore f_i is infected. We proceed in a similar way for the vertex T_i.

The opposite direction is clear, as if t_i or f_i get infected, then in the next round they spread the infection to their neighbours and T_i and F_i immediately become infected due to neighbours on (T_i, F_i)-path.

If u is part of the path connecting T_i and F_i inside the variable gadget X_i, then the vertices of this gadget become infected if and only if both t_i and f_i become infected. However, this can happen only if all the clause gadgets connected to this X_i get infected due to some other variable gadgets. Translated into the language of logic, all clauses in which the variable x_i appears are already satisfied, and it does not matter what value the variable x_i obtains. Therefore, we can select T_i or F_i arbitrarily to S.

Finally, we show that if u is neither in $\{T_i, F_i\}$ nor a member of the path connecting T_i and F_i within X_i, then we can replace u with either T_i or F_i in S to obtain another target set S'. Let $u \in V(X_i)$ be such a vertex. If $u \in N(F_i) \cup \{f_i\}$, then $S \setminus \{u\} \cup \{F_i\}$ is indeed a target set, as it infects the same vertices as u and F_i is even earlier infected. We proceed analogously for the remaining possibilities of u, but instead of replacing with F_i, we replace with T_i.   ◀

Now, we construct a satisfying assignment π for φ. Let X_i be a variable gadget, and let $u = S \cap \{T_i, F_i\}$. If $u = T_i$, then we set $\pi(x_i) = 1$, and if $u = F_i$, then we set $\pi(x_i) = 0$. Note that such S always exists due to Claim 3.

It remains to state that the presented reduction can be done in polynomial time and that the problem is clearly in NP as we can guess the target set S of size k and verify the solution by simulation of the activation process.   □

As stated before, grid graphs are a subfamily of unit disk graphs. Hence, we obtain NP-hardness for unit disks as a corollary of Theorem 5.

Corollary 1. TARGET SET SELECTION *is NP-complete even when the underlying graph is a unit disk graph and all thresholds are at most* 2.

5   Majority Thresholds

The last natural restriction of the threshold function, which is heavily studied in the literature, is the case of majority thresholds, that is, for every $v \in V$ we have $t(v) = \lceil \deg(v)/2 \rceil$.

Theorem 6. TARGET SET SELECTION *is NP-complete even when the underlying graph is a grid graph and all agents have majority thresholds.*

Proof (sketch). The reduction is weakly the same as in the proof of Theorem 5, hence, we highlight only the main differences in the construction of our gadgets. The only problematic part of our gadgets are pendant vertices with threshold equal to 2. Hence, we replace these with an induced path on three vertices called a *cherry* as follows.

In particular, for every clause gadget Y_i, $i \in [m]$, the vertex y_i already has majority thresholds. For the vertex g_i, we add two pendant vertices g_i^1 and g_i^2, left $t(g_i) = 2$, and set $t(g_i^1) = t(g_i^2) = 1$. We call a triple (g_i^1, g_i, g_i^2) a *cherry*. Note that all vertices of the clause gadget have majority thresholds.

For variable gadgets, we do the same adjustment as in the case of clause gadgets. Let u be a pendant vertex of some variable gadget. We add two pendant vertices u_1 and u_2 connected to u with $t(u_1) = t(u_2) = 1$. Every triple (u_1, u, u_2) is another cherry.

It is not hard to see that such a change does change neither the required budget k nor the correctness of the construction.

Claim 4. Let (x, y, z) be a cherry and S be a target-set for an instance $\mathcal{I}$. Then at least one vertex of this cherry is in S.

Proof. Suppose that no vertex of this cherry is part of S. The only vertex with a neighbour outside the cherry is y and we have $t(y) = 2$. However, there is only one neighbour of y outside the cherry; hence y never gets infected, which is a contradiction to the fact that S is a target set.   ◄

Claim 5. Let (x, y, z) be a cherry and S be a target set such that $y \notin S$. Then there is a target set S' such that $y \in S'$ with $|S'| \leq |S|$.

Proof. Suppose that $x, z \in S$. We set $S' = (S \setminus \{x, z\}) \cup \{y\}$. Observe that the process started with S infects y in the second round. If we start the process with S', then y is initially infected and both x, z will be infected in the second round. Moreover, $|S'| = |S| - 1$.

The remaining case is when only one of x, z is part of S. Without loss of generality, let $x \in S$. Since S is a target set, the vertex y becomes infected due to x and the neighbour outside the cherry. Hence, by setting $S' = (S \setminus \{x\}) \cup \{y\}$ we also have a target set of the same size.   ◄

We note that the graph remains planar and $\Delta(G) = 4$, and hence the graph G can be turned into a grid graph using the theorem of Valiant [34].   □

Since the family of grid graphs is a subfamily of unit disk graphs, we directly obtain the hardness even in the case of unit disk graphs.

Corollary 2. TARGET SET SELECTION *is NP-complete even when the underlying graph is a unit disk graph and all agents have majority thresholds.*

6   Conclusions

In this paper, we showed that TARGET SET SELECTION is computationally hard even on very simple geometric graph classes such as unit disk graphs. For more restricted graph families, we showed that TSS is polynomial-time solvable on interval graphs and grid graphs if the threshold function is unanimous, however, in case of grid graphs, we were not able to identify any other tractable restriction of threshold function. In particular, we showed that TSS is NP-complete even if restricted to constant-bounded and majority thresholds.

One of the questions that remains open even after this paper is related to the computational complexity of the TARGET SET SELECTION problem on interval graphs. It was known that TSS is polynomial-time solvable on interval graphs if the threshold function is bounded by a constant and we showed that the same holds even in the case of unanimous thresholds. We conjecture that the problem should be tractable with majority thresholds; however, we are more sceptical in the case of general thresholds.

Acknowledgments. The authors acknowledge the support of the Czech Science Foundation Grant No. 22-19557S. MD and ŠS were additionally supported by the Grant Agency of the Czech Technical University in Prague, grant No. SGS23/205/OHK3/3T/18. MD was also supported by the Student Summer Research Program 2021 of FIT CTU in Prague.

References

1. Araújo, R.T., Sampaio, R.M., dos Santos, V.F., Szwarcfiter, J.L.: The convexity of induced paths of order three and applications: complexity aspects. Discret. Appl. Math. **237**, 33–42 (2018). https://doi.org/10.1016/j.dam.2017.11.007
2. Ben-Zwi, O., Hermelin, D., Lokshtanov, D., Newman, I.: Treewidth governs the complexity of target set selection. Discrete Optim. **8**(1), 87–96 (2011). https://doi.org/10.1016/j.disopt.2010.09.007
3. Bessy, S., Ehard, S., Penso, L.D., Rautenbach, D.: Dynamic monopolies for interval graphs with bounded thresholds. Discret. Appl. Math. **260**, 256–261 (2019). https://doi.org/10.1016/j.dam.2019.01.022
4. Blažej, V., Knop, D., Schierreich, Š.: Controlling the spread of two secrets in diverse social networks (student abstract). In: AAAI 2022, pp. 12919–12920. AAAI Press (2022). https://ojs.aaai.org/index.php/AAAI/article/view/21596
5. Breu, H., Kirkpatrick, D.G.: Unit disk graph recognition is NP-hard. Comput. Geom. **9**(1), 3–24 (1998). https://doi.org/10.1016/S0925-7721(97)00014-X
6. Chen, N.: On the approximability of influence in social networks. SIAM J. Discrete Math. **23**(3), 1400–1415 (2009). https://doi.org/10.1137/08073617X
7. Chiang, C.Y., Huang, L.H., Li, B.J., Wu, J., Yeh, H.G.: Some results on the target set selection problem. J. Comb. Optim. **25**(4), 702–715 (2013). https://doi.org/10.1007/s10878-012-9518-3
8. Chopin, M., Nichterlein, A., Niedermeier, R., Weller, M.: Constant thresholds can make target set selection tractable. Theory Comput. Syst. **55**(1), 61–83 (2013). https://doi.org/10.1007/s00224-013-9499-3
9. Cicalese, F., Cordasco, G., Gargano, L., Milanič, M., Vaccaro, U.: Latency-bounded target set selection in social networks. Theor. Comput. Sci. **535**, 1–15 (2014). https://doi.org/10.1016/j.tcs.2014.02.027
10. Clark, B.N., Colbourn, C.J., Johnson, D.S.: Unit disk graphs. Discrete Math. **86**(1), 165–177 (1990). https://doi.org/10.1016/0012-365X(90)90358-O
11. Dahlhaus, E., Johnson, D.S., Papadimitriou, C.H., Seymour, P.D., Yannakakis, M.: The complexity of multiterminal cuts. SIAM J. Comput. **23**(4), 864–894 (1994). https://doi.org/10.1137/S0097539792225297
12. Dimitrov, D.: The social choice approach to group identification. In: Herrera-Viedma, E., García-Lapresta, J.L., Kacprzyk, J., Fedrizzi, M., Nurmi, H., Zadrożny, S. (eds.) Consensual Processes, pp. 123–134. Springer, Heidelberg (2011). https://doi.org/10.1007/978-3-642-20533-0_7
13. Dreyer, P.A., Roberts, F.S.: Irreversible k-threshold processes: graph-theoretical threshold models of the spread of disease and of opinion. Discret. Appl. Math. **157**(7), 1615–1627 (2009). https://doi.org/10.1016/j.dam.2008.09.012
14. Dvořák, P., Knop, D., Toufar, T.: Target set selection in dense graph classes. SIAM J. Discrete Math. **36**(1), 536–572 (2022). https://doi.org/10.1137/20M1337624
15. Erdélyi, G., Reger, C., Yang, Y.: The complexity of bribery and control in group identification. Auton. Agent. Multi-Agent Syst. **34**(1), 1–31 (2019). https://doi.org/10.1007/s10458-019-09427-9
16. Farber, M.: Independent domination in chordal graphs. Oper. Res. Lett. **1**(4), 134–138 (1982). https://doi.org/10.1016/0167-6377(82)90015-3
17. Finbow, S., MacGillivray, G.: The firefighter problem: a survey of results, directions and questions. Australas. J. Comb. **43**, 57–77 (2009)
18. Fleischner, H., Sabidussi, G., Sarvanov, V.I.: Maximum independent sets in 3- and 4-regular hamiltonian graphs. Discrete Math. **310**(20), 2742–2749 (2010). https://doi.org/10.1016/j.disc.2010.05.028

19. Fomin, F.V., Heggernes, P., van Leeuwen, E.J.: The firefighter problem on graph classes. Theor. Comput. Sci. **613**, 38–50 (2016). https://doi.org/10.1016/j.tcs.2015.11.024
20. Gallai, T.: Über extreme punkt- und kantenmengen. Ann. Univ. Sci. Budapest **2**, 133–138 (1959)
21. Hartmann, T.A.: Target set selection parameterized by clique-width and maximum threshold. In: Tjoa, A.M., Bellatreche, L., Biffl, S., van Leeuwen, J., Wiedermann, J. (eds.) SOFSEM 2018. LNCS, vol. 10706, pp. 137–149. Springer, Cham (2018). https://doi.org/10.1007/978-3-319-73117-9_10
22. Hartnell, B.L.: Firefighter! an application of domination (1995)
23. Hliněný, P., Kratochvíl, J.: Representing graphs by disks and balls (a survey of recognition-complexity results). Discrete Math. **229**(1), 101–124 (2001). https://doi.org/10.1016/S0012-365X(00)00204-1
24. Huson, M.L., Sen, A.: Broadcast scheduling algorithms for radio networks. In: MILCOM 1995, vol. 2, pp. 647–651 (1995). https://doi.org/10.1109/MILCOM.1995.483546
25. Kang, R.J., Müller, T.: Sphere and dot product representations of graphs. Discrete Comput. Geom. **47**(3), 548–568 (2012). https://doi.org/10.1007/s00454-012-9394-8
26. Kasher, A., Rubinstein, A.: On the question "who is a J?" a social choice approach. Logique et Analyse **40**(160), 385–395 (1997)
27. Kempe, D., Kleinberg, J., Tardos, É.: Maximizing the spread of influence through a social network. Theory Comput. **11**(4), 105–147 (2015). https://doi.org/10.4086/toc.2015.v011a004
28. Mathieson, L.: The parameterized complexity of editing graphs for bounded degeneracy. Theor. Comput. Sci. **411**(34–36), 3181–3187 (2010). https://doi.org/10.1016/j.tcs.2010.05.015
29. Matsui, T.: Approximation algorithms for maximum independent set problems and fractional coloring problems on unit disk graphs. In: Akiyama, J., Kano, M., Urabe, M. (eds.) JCDCG 1998. LNCS, vol. 1763, pp. 194–200. Springer, Heidelberg (2000). https://doi.org/10.1007/978-3-540-46515-7_16
30. Nichterlein, A., Niedermeier, R., Uhlmann, J., Weller, M.: On tractable cases of target set selection. Soc. Netw. Anal. Min. **3**(2), 233–256 (2013). https://doi.org/10.1007/s13278-012-0067-7
31. Peleg, D.: Majority voting, coalitions and monopolies in graphs. In: SIROCCO 1996, pp. 152–169. Carleton Scientific (1996)
32. Reddy, T.V.T., Krishna, D.S., Rangan, C.P.: Variants of spreading messages. In: Rahman, M.S., Fujita, S. (eds.) WALCOM 2010. LNCS, vol. 5942, pp. 240–251. Springer, Heidelberg (2010). https://doi.org/10.1007/978-3-642-11440-3_22
33. Richardson, M., Domingos, P.: Mining knowledge-sharing sites for viral marketing. In: KDD 2002, pp. 61–70. ACM, New York (2002). https://doi.org/10.1145/775047.775057
34. Valiant, L.G.: Universality considerations in VLSI circuits. IEEE Trans. Comput. **30**(02), 135–140 (1981). https://doi.org/10.1109/TC.1981.6312176
35. Yang, Y., Dimitrov, D.: How hard is it to control a group? Auton. Agent. Multi-Agent Syst. **32**(5), 672–692 (2018). https://doi.org/10.1007/s10458-018-9392-1

Multilayer Hypergraph Clustering Using the Aggregate Similarity Matrix

Kalle Alaluusua[1($\boxtimes$)], Konstantin Avrachenkov[2], B. R. Vinay Kumar[2], and Lasse Leskelä[1]

[1] Aalto University, Espoo, Finland
{kalle.alaluusua,lasse.leskela}@aalto.fi
[2] INRIA, Sophia Antipolis, Valbonne, France
{k.avrachenkov,vinay-kumar.bindiganavile-ramadas}@inria.fr

Abstract. We consider the community recovery problem on a multilayer variant of the hypergraph stochastic block model (HSBM). Each layer is associated with an independent realization of a d-uniform HSBM on N vertices. Given the similarity matrix containing the aggregated number of hyperedges incident to each pair of vertices, the goal is to obtain a partition of the N vertices into disjoint communities. In this work, we investigate a semidefinite programming (SDP) approach and obtain information–theoretic conditions on the model parameters that guarantee exact recovery both in the assortative and the disassortative cases.

Keywords: hypergraph SBM · community detection · semidefinite programming · multilayer · clustering · planted partition

1 Introduction

Traditional network data are observed as interactions between node pairs, represented as a graph, or equivalently as an adjacency matrix. More refined forms of network data may involve multiple types of higher-order interactions simultaneously involving multiple nodes. Such a data set can be viewed as a binary array $A_e^{(m)}$ indexed by node sets e and positive integers m so that $A_e^{(m)} = 1$ indicates that an interaction of type m occurs among node set e. The array can also be viewed as a multilayer hypergraph where the entries of the array indicate the presence of hyperedge e in the m-th layer. Further, the entries could also depend on the communities of the underlying nodes. Stochastic block models (SBMs) are a popular choice for generative models with a community structure for such applications. Hypergraph stochastic block models (HSBMs) introduce hyperedges into SBMs, thus extending their modelling capabilities. In the following, we provide two examples to illustrate the relevance of the model discussed in this work

Supported by the Vilho, Yrjö and Kalle Väisälä Foundation of the Finnish Academy of Science and Letters, and the French government through the RISE Academy of UCA[JEDI] Investments in the Future project managed by the National Research Agency (ANR) with the reference number ANR-15-IDEX-0001.

M. Dewar et al. (Eds.): WAW 2023, LNCS 13894, pp. 83–98, 2023.
https://doi.org/10.1007/978-3-031-32296-9_6

1. **Table reservations at restaurants**: Consider M restaurants each of which have tables that can accommodate d people. A population of N individuals is divided into two different communities; those who prefer vegan and those who do not. A hyperedge links d individuals with one of the M restaurants indicating that the corresponding d individuals made a reservation at the restaurant at some point of time. Thus, each of the M restaurants correspond to a layer. Naturally, individuals who prefer vegan dishes gravitate towards those restaurants which have more vegan options. But since the choice of a restaurant is made by a group of d individuals who can belong to either community, the probability that a restaurant is visited will depend on the number of individuals in each of the communities. In other words, a subset of d individuals form a hyperedge which is present with probability dependent on the communities of the d individuals.

2. **Processor sharing**: N tasks are assigned to M heterogeneous servers each of which has several d-core processors that can process d tasks at a time. A subset of d tasks constitute a hyperedge, and is assigned to a processor based on the priorities (high or low) of each of the tasks and the computation capability of the server.

We will next describe a generative model of a hypergraph with $N \geq 1$ nodes and $M \geq 1$ layers, where each hyperedge has size $d \geq 2$. We take M and d as constants independent of N. The set of nodes is divided into two communities of equal sizes (we assume N is even), and the resulting community structure, denoted by $\sigma^{(N)}$, is uniformly distributed on the set $\{(\sigma_1, \sigma_2, \cdots, \sigma_N) \in \{-1,+1\}^N : |\{i : \sigma_i = +1\}| = |\{i : \sigma_i = -1\}|\}$. The community profile of a node set e is defined as a vector $\tau(e) = (\tau_{-1}(e), \tau_{+1}(e))$, where $\tau_k(e)$ is equal to the number of nodes in e with community membership k. We will then sample a multilayer hypergraph on node set $[N] = \{1, \ldots, N\}$ so that each node set $e \subset [N]$ having size d and community profile $\tau(e) = t$ is linked by a hyperedge in layer m with probability

$$p_t^{(m)} = \frac{\alpha_t^{(m)} \log N}{\binom{N-1}{d-1}} \wedge 1, \tag{1}$$

independently of other node sets and layers. Here $\alpha_t^{(m)} > 0$ is a real number independent of N. This scaling of the hyperedge probabilities ensures that the expected average degree of each node is $\Theta(\log N)$. References [2,10,26] show that the phase-transition for exact recovery occurs in this regime, and this regime is also critical for connectivity in general hypergraph models [7].

The d-uniform multilayer hypergraph can be represented as a binary array $A = (A_e^{(m)})$ in which the entries are mutually independent Bernoulli random variables. The event $\{A_e^{(m)} = 1\}$ has probability $p_t^{(m)}$ when $\tau(e) = t$. To indicate that the pair $(\sigma, A) = (\sigma^{(N)}, A^{(N)})$ is sampled from the model, we abbreviate $(\sigma, A) \sim \text{HSBM}(N, M, d, (\alpha_t^{(m)}))$. We will focus on a *symmetric* model in which

$$\alpha_{(r,d-r)}^{(m)} = \alpha_{(d-r,r)}^{(m)}$$

for all $0 \leq r \leq d$ and all m. This means that the presence of the hyperedge depends only on the number of nodes of each community rather than the community label.

The problem of community detection is to output an estimate $\hat{\sigma}^{(N)}$ of the underlying node communities. The estimate is said to achieve *exact recovery* if,

$$\lim_{N \to \infty} \mathbb{P}\left(\hat{\sigma}^{(N)} \in \{\pm \sigma^{(N)}\}\right) = 1. \tag{2}$$

where $\mathbb{P}$ is the probability measure associated with $\text{HSBM}(N, M, d, (\alpha_t^{(m)}))$, a generative model for the community structure and the observations.

In this work, the main focus is to study the community recovery problem based on a layer-aggregated similarity matrix $W_{ij} = \sum_m W_{ij}^{(m)}$ where $(W_{ij}^{(m)}) =: W^{(m)}$ is a zero-diagonal matrix with off-diagonal entries

$$W_{ij}^{(m)} = \sum_{e:e \ni i,j} A_e^{(m)}$$

counting the number of hyperedges in layer m incident to nodes i and j. Community recovery based on the similarity matrix W instead of the full data set A is motivated by two aspects: privacy and computational tractability. For example, in the application of table reservations at restaurants, providing the full hypergraph could reveal the frequency a particular individual visits a restaurant. This could violate the privacy of the individual. On the other hand, providing the similarity matrix obfuscates such individual information, since information regarding the restaurants that are visited is not revealed. Additionally, the similarity matrix provides a compact matrix representation of the hypergraph that is easier to manipulate using matrix algebra. Nevertheless, it is clear that the similarity matrix contains less information than the complete hypergraph.

In this work, we investigate a semidefinite programming (SDP) approach for solving the community detection problem using W. Our main result in Sect. 3 provides an information quantity that characterizes the performance of the SDP approach. To be more specific, it gives a sufficient condition relating the different parameters of the model for exact recovery using the SDP technique.

The paper is organized as follows. In Sect. 2, we discuss recent literature in the area and highlight the key differences in our work. In Sect. 3, we describe the semidefinite programming algorithm and state our main result concerning the information theoretic conditions on the parameters of the model that guarantee exact recovery. Section 4 contains some simulation results of our algorithm on synthetic data generated using the HSBM model. The proofs of our results are provided in Sect. 5. Section 6 concludes the paper and provides some future directions.

2   Related Work

The usual paradigm for theoretical research in community detection is to first propose a generative model that captures the application (data) as a graph or a network, followed by the analysis of a clustering algorithm for the proposed model. Community recovery algorithms on stochastic block models (SBMs) have attracted considerable attention in the past. A comprehensive survey of the field is provided in [1], see also a review of recent results on graph clustering in [5]. Our interest in this work is on multilayer hypergraphs. We first provide a brief survey of recent work in clustering on multilayer networks followed by hypergraph SBMs.

A seminal work for multilayer networks is the review article [27]. Subsequently, in [34], the authors consider estimating the membership for each individual layer in a multilayer SBM. In the special case that memberships do not change, their method works on the normalized aggregate adjacency matrix. The authors in [31] establish that increasing the number of network layers guarantees consistent community recovery (by a least squares estimator) even when the graphs get sparser. SBMs with general interactions allow an alternate model for multilayer networks. These are studied in [6] where the authors address the community recovery problem using aggregate adjacency matrix as well as the full graph. The authors in [3] study Bayesian community recovery in a regime where both the number of nodes and the number of layers increase.

With regard to literature on hypergraphs, the hypergraph stochastic block model (HSBM) was first introduced by Ghoshdastidar and Dukkipati in [13] to capture higher order interactions. They also show strong consistency of spectral methods in dense uniform hypergraphs. In subsequent works [14–16], they extend their results to partial recovery in sparse non-uniform hypergraphs. Some other works on the spectral algorithm for hypergraphs include [2,33].

The recent work by Zhang and Tan [36] considers the general d-uniform HSBM with multiple communities. They establish a sharp phase transition for exact recovery when the knowledge of the whole hypergraph is given. They recover results from several previous works including [9,10,26,33]. In the process of solving the exact recovery problem, they do show almost exact recovery using only the similarity matrix through a spectral approach. Another general hypergraph model with theoretical guarantees by [37] employs a latent space representation of nodes to cover HSBM, non-uniform hypergraphs, and hypergraphs with heterogeneity among nodes.

Some of the other approaches used in the literature for the community recovery problem on hypergraphs are based on modularity [21–23,28,29], tensor decomposition [24,37], random walk based methods [33,35,38], variational inference [8], and approximate message passing [4,32].

In this work, we investigate the problem of exact recovery on the HSBM through the lens of semidefinite programming (SDP). Our work is closest in spirit to [12] and [26] that discuss the SDP approach. The SDP formulation (described in Sect. 3) arises as a relaxation of the computationally hard procedure of finding a nodes' partition with minimum number of edges crossing it. In [26], the authors show that for a d-uniform homogeneous HSBM with two equal-sized and symmetric communities, exact recovery using the full hypergraph shows a sharp phase transition behavior. They go on to propose a 'truncate-and-relax' algorithm that utilizes the structure of the similarity matrix. An SDP approach then guarantees exact recovery with high probability, albeit in a parameter regime which is slightly sub-optimal. This gap is bridged in [12] who consider the community recovery problem with the knowledge of only the similarity matrix. Below, we highlight the differences from these previous works:

1. The authors in both [12] and [26] consider the homogeneous model in which the hyperedge parameters take just two values corresponding to all nodes of a hyperedge being in the same community, and at least one of them being in a different community. Related works with the same assumption include [2,11,30]. In this work, we allow for hyperedge parameters to depend on the number of nodes of each community in the hyperedge resulting in an inhomogeneous HSBM as in [36], albeit with the symmetric assumption. A similar assumption is present in other works such as [13,15] as well.
2. Much of the earlier work assumes that the data is *assortative* or *homophilic*, i.e. nodes in the same community are more likely to be adjacent to each other than to nodes in different communities. Our results incorporate the *disassortative* or *heterophilic* case where the opposite is true. This could be of interest for some applications: reputation of a research institute is partly assessed based on the amount of collaboration with experts from external institutions (see e.g. [18]); so, one might expect certain research networks to be disassortative.
3. Our model targets multilayer HSBMs that can be seen as a generalization of previous models. Moreover, these layers could individually be assortative or disassortative which could then capture a plethora of applications.

3   Algorithm and Main Results

A first approach to obtain an estimate of the node communities given the similarity matrix W is to solve the min-bisection problem:

$$\max \sum_{i,j} W_{ij} x_i x_j \quad \text{subject to } \boldsymbol{x} \in \{\pm 1\}^N, \mathbf{1}^T \boldsymbol{x} = 0. \tag{3}$$

This formulation assumes that the data is assortative. In the disassortative case the opposite is true, and we replace the maximization in (3) with minimization, or equivalently change the sign of the objective function. However, the min-bisection problem is known to be NP-hard in general (see [25]), which is why [26] introduces a semidefinite programming (SDP) relaxation of (3). Algorithm 1 introduces an additional input $s \in \{\pm 1\}$ and generalizes their relaxation to both assortative ($s = +1$) and disassortative ($s = -1$) cases.

Remark 1. An alternate approach in the disassortative case is to consider the complement of the hypergraph, which is assortative. A similarity matrix of the complement is given by $\binom{N-2}{d-2}(\mathbf{1}\mathbf{1}^T - I) - \boldsymbol{W}$. However, owing to our scaling assumption in (1), the resulting similarity matrix of the complement is no longer in the same regime, and requires a different approach to analyze.

Algorithm 1

Input: $N \times N$ similarity matrix $\boldsymbol{W}$ and $s \in \{\pm 1\}$.
Output: Community estimate $\hat{\boldsymbol{\sigma}}$
1: Solve the following optimization problem:

$$
\begin{aligned}
\text{maximize} \quad & \sum_{0 \leq i < j \leq N} s W_{ij} X_{ij} \\
\text{subject to} \quad & \sum_{0 \leq i < j \leq N} X_{ij} = 0, \\
& X_{ii} = 1 \text{ for all } i \in [N] \\
& \boldsymbol{X} \succeq 0.
\end{aligned}
\tag{4}
$$

2: Let $\boldsymbol{X}^* \in \mathbb{R}^{N \times N}$ be the optimal solution of (4) and let it have an eigendecomposition $\boldsymbol{X}^* = \sum_{i=1}^N \lambda_i \boldsymbol{v}_i \boldsymbol{v}_i^{\mathrm{T}}$ with $\lambda_1 \geq \lambda_2 \geq \cdots \geq \lambda_N$.
3: Output $\hat{\boldsymbol{\sigma}} = \mathrm{sgn}(\boldsymbol{v}_1)$

To capture the level of assortativity of our model, we define the following quantity, accordingly referred to as the *assortativity*

$$
\xi := \sum_{m=1}^M \sum_{r=0}^{d-1} \binom{d-1}{r} (d-1-2r) \alpha_{(r,d-r)}^{(m)}.
\tag{5}
$$

The summation over the different layers implies that the full hypergraph can be assortative even if individual layers $\boldsymbol{W}^{(m)}$ are not. Table 1 specifies Formula (5) for selected values of d. The following proposition (proved in Sect. 5.4) states that ξ is a normalized expected difference between the number of hyperedges shared between two nodes when they are of the same community and when they are of different communities.

Proposition 1. *For $i \neq j$, let $w_{\mathrm{in}} = \mathbb{E}[W_{ij}|\sigma_i = \sigma_j]$ and $w_{\mathrm{out}} = \mathbb{E}[W_{ij}|\sigma_i \neq \sigma_j]$. Then,*

$$
w_{\mathrm{in}} - w_{\mathrm{out}} = \frac{\log N}{2^{d-2} N} \xi + o\left(\frac{\log N}{N}\right).
$$

Therefore, a model is said to be assortative if $\xi > 0$, and disassortative if $\xi < 0$.

In order to state our main result concerning the performance of Algorithm 1, we will need the following information quantity

$$
I = \sup_{\lambda \in \mathbb{R}} \sum_{m=1}^M \sum_{r=0}^{d-1} 2^{-(d-1)} \binom{d-1}{r} \alpha_{(r,d-r)}^{(m)} \left(1 - e^{-\lambda(d-1-2r)}\right),
\tag{6}
$$

which captures the propensity of a node to be present in a hyperedge containing more nodes of the same (different) community as itself in the assortative (disassortative) case. To be concise, we omit the dependence on model parameters in the definition of both ξ and I.

Table 1. Assortativity ξ in a d-uniform HSBM with M layers, where we denote $\alpha_t = \sum_{m=1}^{M} \alpha_t^{(m)}$.

d	ξ
2	$\alpha_{(0,2)} - \alpha_{(1,1)}$
3	$2\alpha_{(0,3)} - 2\alpha_{(1,2)}$
4	$3\alpha_{(0,4)} - 3\alpha_{(2,2)}$
5	$4\alpha_{(0,5)} + 4\alpha_{(1,4)} - 8\alpha_{(2,3)}$
6	$5\alpha_{(0,6)} + 10\alpha_{(1,5)} - 5\alpha_{(2,4)} - 10\alpha_{(3,3)}$

We now state the main theorem that characterizes Algorithm 1 and provides a sufficient condition for exact recovery on the aggregate similarity matrix of a symmetric multilayer HSBM using the SDP approach.

Theorem 1. *Suppose* $(\boldsymbol{\sigma}, \boldsymbol{A}) \sim \mathrm{HSBM}(N, M, d, (\alpha_t^{(m)}))$, *and let* $\boldsymbol{W}$ *be the aggregate similarity matrix of* $\boldsymbol{A}$. *When* $I > 1$, *Algorithm 1 with* $\boldsymbol{W}$ *and* $s = \mathrm{sgn}(\xi)$ *as inputs achieves exact recovery as defined in* (2).

The proof of Theorem 1 is provided in Sect. 5. Taking $M = 1$ with parameters

$$\alpha_{(r,d-r)} = \begin{cases} \alpha & \text{for } r = 0 \text{ and } r = d \\ \beta & \text{for } 1 \leq r \leq d-1 \end{cases}$$

reduces to a homogeneous model that has been studied earlier in the assortative case with $\xi = (d-1)(\alpha - \beta) > 0$. In this setting Kim, Bandeira, and Goemans [26] showed that the SDP algorithm does not achieve exact recovery when $I < 1$ and Gaudio and Joshi [12] proved that the SDP algorithm achieves exact recovery when $I > 1$, as conjectured in [26].

4   Numerical Illustrations

We perform numerical simulations to demonstrate the effect of the number of observed hypergraph layers on the classification accuracy of Algorithm 1[1]. The synthetic data is sampled from a 4-uniform $\mathrm{HSBM}(50, M, 4, (\alpha_t^{(m)}))$, with $1 \leq M \leq 3$. We let the hypergraph layers be identically distributed giving $\alpha_t^{(m)} =: \alpha_t$ for all $m \in [M]$. We examine four scenarios: homogeneous and assortative, homogeneous and disassortative, inhomogeneous and assortative, as well as inhomogeneous and disassortative. Table 2 provides the parameter values used in each case, respectively. These values are chosen such that the expected degree, i.e. the number of hyperedges a node is incident to, is the same in both the homogeneous and the inhomogeneous cases. The associated hyperedge probabilities are computed from (1).

---

[1] Source code: https://github.com/kalaluusua/Hypergraph-clustering.git.

Table 2. The columns with numerical entries represent the parameter values (α_t) used in each of the four simulated scenarios.

	Homogeneous		Inhomogeneous	
	Assortat.	Disassort.	Assortat.	Disassort.
$\alpha_{(4,0)}$	18.8	7.3	18.8	4.7
$\alpha_{(3,1)}$	7.3	18.8	9.4	9.4
$\alpha_{(2,2)}$	7.3	18.8	4.7	18.8

To evaluate the accuracy of our estimate given σ, we use the Hubert-Arabie adjusted Rand index (AR) [17,20], which is a measure of similarity between two community assignments. The index is equal to 1 when the assignments are identical, and 0 when they are independent. For each simulated hypergraph, we also compute the classification error (CE), which we define as the fraction of misclassified nodes $N^{-1}\min\{\mathrm{Ham}(\hat{\sigma},\sigma),\mathrm{Ham}(\hat{\sigma},-\sigma)\}$, where Ham denotes the Hamming distance. The results (averaged over five different random seed initializations) are depicted in Table 3.

Table 3. Classification error (CE) and Adjusted Rand index (AR) of the community assignment estimate.

M	Homogeneous						Inhomogeneous					
	$|\xi|$	I	Assortat.		Disassort.		$|\xi|$	I	Assortat.		Disassort.	
			CE	AR	CE	AR			CE	AR	CE	AR
1	34.5	0.58	0.160	0.464	0.184	0.475	42.4	0.41	0.052	0.799	0.012	0.952
2	68.9	1.08	0.024	0.906	0.052	0.800	84.8	0.83	0.008	0.969	0.000	1.000
3	103.4	1.62	0.004	0.984	0.012	0.953	127.2	1.24	0.000	1.000	0.000	1.000

Based on the I-values, we expect that the community detection performance improves as M increases and is the same for the assortative and disassortative cases. This is precisely what Table 3 shows. Moreover, the larger I-values of the homogeneous case lead us to expect an overall better performance in comparison to the inhomogeneous case. Surprisingly, this is not the case. An inspection of the ξ-values reveals a larger level of (dis)assortativity in the inhomogeneous case. In small to moderate hypergraph sizes, we suspect that the level of assortativity may predict the detection performance of Algorithm 1 better than the information-theoretic quantity I, whose effect is more profound in the asymptotic regime.

5   Analysis of the Algorithm

In this section, we provide a detailed proof of Theorem 1. We follow the procedure of [12,19,26] to analyze the SDP framework, and extend it to a more general model $\text{HSBM}(N, M, d, (\alpha_t^{(m)}))$ that addresses multiple layers, disassortativity, and (symmetric) inhomogeneity.

An outline of this section is as follows. Section 5.1 constructs a dual certificate strategy to solve the SDP in (4) and specializes it to the assortative and disassortative cases. Bounds on certain quantities that arise as part of this strategy are provided in Sects. 5.2 and 5.3. In Sect. 5.4, we comment on the assortative/disassortative nature of the model and its manifestation in our analysis. Finally, Sect. 5.5 puts the parts together to complete the proof of Theorem 1.

5.1 SDP Analysis

To begin, we state a sufficient condition for optimality of Algorithm 1. This is a corollary of [12, Lemma 2.2] that asserts strong duality for (4) with $s = 1$.

Lemma 1. *Fix $s \in \{\pm 1\}$. Suppose there is a diagonal matrix $D \in \mathbb{R}^{N \times N}$ and $\nu \in \mathbb{R}$ such that the matrix $S := D + \nu 11^T - sW$ is positive semidefinite, its second smallest eigenvalue $\lambda_{N-1}(S)$ is strictly positive, and $S\sigma = 0$, then $X^* = \sigma\sigma^T$ is the unique optimal solution to (4) (with the same s).*

For the $\text{HSBM}(N, M, d, (\alpha_t^{(m)}))$ model with node communities σ and the aggregate similarity matrix W, define

$$D_{ii} := s \sum_j W_{ij}\sigma_i\sigma_j, \tag{7}$$

where $s = \text{sgn}(\xi)$. With $D = \text{diag}(D_{ii})$, it is easy to verify that $S\sigma = 0$ and, therefore, it suffices to show that

$$\mathbb{P}\left(\inf_{x \perp \sigma : \|x\|_2 = 1} x^T S x > 0 \right) = 1 - o(1) \tag{8}$$

for Lemma 1 to hold. Note that $S\sigma = 0$ and (8) together ensure that the kernel of S is the line spanned by σ and $\lambda_1(S) \geq \cdots \geq \lambda_{N-1}(S) > 0$ with high probability. This in particular implies that S is positive semidefinite and its second smallest eigenvalue $\lambda_{N-1}(S)$ is strictly positive, with high probability. Using a similar methodology as in [19, Theorem 2], we obtain the following complementary lemmas for the assortative and disassortative cases, respectively.

Lemma 2. *Let $\xi > 0$. With D defined via (7) with $s = \text{sgn}(\xi) = +1$ and $S := D + 11^T - W$, for all $x \perp \sigma$ such that $\|x\|_2 = 1$, we have*

$$x^T S x \geq \min_i D_{ii} - \|W - \mathbb{E}W\|_2,$$

where $\mathbb{E}W$ is the expected aggregate similarity matrix conditioned on σ.

Proof. The expected similarity matrix for a symmetric HSBM admits the following rank-2 decomposition:

$$\mathbb{E}W = \left(\frac{w_{\text{in}} + w_{\text{out}}}{2}\right) 11^T + \left(\frac{w_{\text{in}} - w_{\text{out}}}{2}\right) \sigma\sigma^T - w_{\text{in}}I,$$

where $w_{\text{in}} = \mathbb{E}[W_{ij}|\sigma_i = \sigma_j]$, $w_{\text{out}} = \mathbb{E}[W_{ij}|\sigma_i \neq \sigma_j]$ and I is the $N \times N$ identity matrix. We can then write

$$
\begin{aligned}
\boldsymbol{x}^T \boldsymbol{S} \boldsymbol{x} &= \boldsymbol{x}^T \boldsymbol{D} \boldsymbol{x} + \left(\boldsymbol{1}^T \boldsymbol{x}\right)^2 - \boldsymbol{x}^T (\boldsymbol{W} - \mathbb{E}\boldsymbol{W})\boldsymbol{x} - \boldsymbol{x}^T \mathbb{E}\boldsymbol{W} \boldsymbol{x} \\
&= \boldsymbol{x}^T \boldsymbol{D} \boldsymbol{x} + \left(\boldsymbol{1}^T \boldsymbol{x}\right)^2 - \boldsymbol{x}^T (\boldsymbol{W} - \mathbb{E}\boldsymbol{W})\boldsymbol{x} \\
&\quad - \left(\frac{w_{\text{in}} + w_{\text{out}}}{2}\right) \left(\boldsymbol{1}^T \boldsymbol{x}\right)^2 - \left(\frac{w_{\text{in}} - w_{\text{out}}}{2}\right) \left(\boldsymbol{\sigma}^T \boldsymbol{x}\right)^2 + w_{\text{in}} \|\boldsymbol{x}\|_2^2.
\end{aligned}
$$

Because of the definition of the spectral norm and the facts that $\boldsymbol{x} \perp \boldsymbol{\sigma}$, and $w_{\text{in}}, w_{\text{out}} = \Theta(\frac{\log N}{N})$ as shown in the proof of Proposition 1 (Sect. 5.4), we obtain

$$
\begin{aligned}
\boldsymbol{x}^T \boldsymbol{S} \boldsymbol{x} &\geq \left(\min_i D_{ii}\right) \|\boldsymbol{x}\|_2^2 + \left(\boldsymbol{1}^T \boldsymbol{x}\right)^2 \left(1 - \frac{w_{\text{in}} + w_{\text{out}}}{2}\right) - \boldsymbol{x}^T (\boldsymbol{W} - \mathbb{E}\boldsymbol{W})\boldsymbol{x} \\
&\geq \min_i D_{ii} - \|\boldsymbol{W} - \mathbb{E}\boldsymbol{W}\|_2,
\end{aligned}
$$

which proves the lemma.

Lemma 3. *Let $\xi < 0$. With $\boldsymbol{D}$ defined via (7) with $s = \operatorname{sgn}(\xi) = -1$ and $\boldsymbol{S} := \boldsymbol{D} + \boldsymbol{W}$, for all $\boldsymbol{x} \perp \boldsymbol{\sigma}$ such that $\|\boldsymbol{x}\|_2 = 1$, we have*

$$
\boldsymbol{x}^T \boldsymbol{S} \boldsymbol{x} \geq \min_i D_{ii} - \|\boldsymbol{W} - \mathbb{E}\boldsymbol{W}\|_2 - w_{\text{in}}.
$$

Proof. The claim follows from applying the techniques from the proof of Lemma 2 on

$$
\boldsymbol{x}^T \boldsymbol{S} \boldsymbol{x} = \boldsymbol{x}^T \boldsymbol{D} \boldsymbol{x} - \boldsymbol{x}^T (\mathbb{E}\boldsymbol{W} - \boldsymbol{W})\boldsymbol{x} + \boldsymbol{x}^T \mathbb{E}\boldsymbol{W} \boldsymbol{x}.
$$

5.2   Upper Bound on $\|\boldsymbol{W} - \mathbb{E}\boldsymbol{W}\|_2$

Let $\mathcal{E}$ be the set of all node sets (hyperedges) $e \subset [N]$ having size d. We denote by $([N], (f_e)_{e \in \mathcal{E}})$ a weighted d-uniform hypergraph with edge weights (f_e), whose similarity matrix is a zero-diagonal matrix with off-diagonal entries (i, j) given by $\sum_{e: e \ni i, j} f_e$.

Lemma 4 (Theorem 4, [30]). *Let $G = ([N], (f_e)_{e \in \mathcal{E}})$, where a random weight $f_e \in [0, 1]$ is independently assigned to each hyperedge $e \in \mathcal{E}$. Denote by $\boldsymbol{W}_f$ the similarity matrix of G. Assume that $\max_{e \in \mathcal{E}} \mathbb{E}[f_e] \leq \frac{c_0 \log N}{\binom{N-1}{d-1}}$. Then there exists a constant $C = C(d, c_0) > 0$ such that*

$$
\mathbb{P}\left(\|\boldsymbol{W}_f - \mathbb{E}\boldsymbol{W}_f\|_2 \leq C\sqrt{\log N}\right) \geq 1 - O(N^{-11}).
$$

To apply Lemma 4, for each $e \in \mathcal{E}$ we define $f_e := M^{-1} \sum_m A_e^{(m)} \in [0, 1]$ as the fraction of layers containing hyperedge e in $(\boldsymbol{\sigma}, \boldsymbol{A}) \sim \operatorname{HSBM}(N, M, d, (\alpha_t^{(m)}))$

over the M layers. Then (f_e) are independent by the independence of $(A_e^{(m)})$, and $\max_{e \in \mathcal{E}} \mathbb{E}[f_e] \leq \frac{\alpha^* \log N}{\binom{N-1}{d-1}}$, where $\alpha^* = \max_{t,m} \alpha_t^{(m)}$. By definition of f_e, the similarity matrix of the weighted hypergraph $([N], (f_e)_{e \in \mathcal{E}})$ is equal to $\boldsymbol{W}/M$, where $\boldsymbol{W}$ is the similarity matrix of $\boldsymbol{A}$. Finally, by Lemma 4 and the definition of the spectral norm,

$$\mathbb{P}\left(\|\boldsymbol{W} - \mathbb{E}\boldsymbol{W}\|_2 \leq CM\sqrt{\log N}\right) \geq 1 - O(N^{-11}).$$

5.3   Lower Bound on D_{ii}

Lemma 5. *Let $I > 1$. Then there exists a constant $\epsilon > 0$ dependent on model parameters such that for all $i \in [N]$,*

$$\mathbb{P}(D_{ii} \leq \epsilon \log N) = o(N^{-1}).$$

Proof. We can write D_{ii} in (7) as

$$D_{ii} = s \sum_m \sum_{j:j \neq i} \sum_{e \in \mathcal{E}:e \ni i,j} A_e^{(m)} \sigma_i \sigma_j = s \sum_m \sum_{e \in \mathcal{E}:e \ni i} A_e^{(m)} \sum_{j \in e \setminus \{i\}} \sigma_i \sigma_j.$$

We will split the sum on the right based on the community profile of the node set $e \setminus \{i\}$. Denote by $\mathcal{T}_{d-1}$ the set of vectors $\boldsymbol{t} = (t_{-1}, t_{+1})$ with nonnegative integer-valued coordinates summing up to $t_{-1} + t_{+1} = d - 1$. For each $\boldsymbol{t} \in \mathcal{T}_{d-1}$, denote by $\mathcal{E}_{i,t}$ the collection of node sets e of size d such that $e \ni i$ and such that the number of nodes $j \in e \setminus \{i\}$ with community membership $\sigma_j = k$ equals t_k for $k = \{-1, +1\}$. Then for any node i with block membership $\sigma_i = k$ and any $e \in \mathcal{E}_{i,t}$,

$$\sum_{j \in e \setminus \{i\}} \sigma_i \sigma_j = t_k - t_{-k}.$$

Therefore, for any i with block membership $\sigma_i = k$, we find that

$$D_{ii} = s \sum_m \sum_{\boldsymbol{t} \in \mathcal{T}_{d-1}} \sum_{e \in \mathcal{E}_{i,t}} A_e^{(m)}(t_k - t_{-k}) = s \sum_m \sum_{\boldsymbol{t} \in \mathcal{T}_{d-1}} (t_k - t_{-k}) Y_{i,t}^{(m)}, \quad (9)$$

where $Y_{i,t}^{(m)} = \sum_{e \in \mathcal{E}_{i,t}} A_e^{(m)}$ equals the number of hyperedges e in layer m that contain i and for which the i-excluded community profile equals $\boldsymbol{t} \in \mathcal{T}_{d-1}$. For any such e, the full community profile equals $\boldsymbol{t} + \boldsymbol{e}_k$, where $\boldsymbol{e}_k$ is a basis vector for the coordinate $k \in \{-1, 1\}$. Furthermore, the size of the set $\mathcal{E}_{i,t}$ equals $R_{k,t} :=$ $|\mathcal{E}_{i,t}| = \binom{\frac{N}{2}-1}{t_k}\binom{\frac{N}{2}}{t_{-k}}$. It follows that the random variables $Y_{i,t}^{(m)}$ are mutually independent and binomially distributed according to $Y_{i,t}^{(m)} \sim \text{Bin}(R_{k,t}, p_{t+e_k}^{(m)})$. Fix $\lambda \geq 0$. By independence and the inequality $1 - x \leq e^x$, we find that the moment-generating function of D_{ii} is bounded by

$$\mathbb{E}\left[e^{-\lambda D_{ii}}\right] = \prod_m \prod_{t\in T_{d-1}} \mathbb{E}\left[e^{-s\lambda(t_k-t_{-k})Y_{i,t}^{(m)}}\right]$$

$$= \prod_m \prod_{t\in T_{d-1}} \left[1 - p_{t+e_k}^{(m)}\left(1 - e^{-s\lambda(t_k-t_{-k})}\right)\right]^{R_{k,t}}$$

$$\leq \prod_{t\in T_{d-1}} \prod_m \exp\left[-R_{k,t}p_{t+e_k}^{(m)}\left(1 - e^{-s\lambda(t_k-t_{-k})}\right)\right].$$

Using the bounds $\left(1 - \frac{i}{n}\right)^j \frac{n^j}{j!} \leq \binom{n}{j} \leq \frac{n^j}{j!}$ and the scaling assumption (1), we find that

$$R_{k,t}p_{t+e_k}^{(m)} = (1+o(1))2^{-(d-1)}\binom{d-1}{t_{-k}}\alpha_{t+e_k}^{(m)}\log N. \tag{10}$$

We conclude that

$$\mathbb{E}\left[e^{-\lambda D_{ii}}\right] \leq e^{-(1+o(1))\psi_k(s\lambda)\log N},$$

where, for $x \in \mathbb{R}$,

$$\psi_k(x) := 2^{-(d-1)}\sum_m \sum_{t\in T_{d-1}}\binom{d-1}{t_{-k}}\alpha_{t+e_k}^{(m)}\left(1 - e^{-x(t_k-t_{-k})}\right).$$

For the inner summation, taking $t_{-k} = r$, we have that $t_k = d - 1 - r$ and $\alpha_{t+e_k} = \alpha_{(r,d-r)} = \alpha_{(d-r,r)}$, thus giving

$$\psi_k(x) = 2^{-(d-1)}\sum_m \sum_{r=0}^{d-1}\binom{d-1}{r}\alpha_{(r,d-r)}^{(m)}\left(1 - e^{-x(d-1-2r)}\right) =: \psi(x).$$

Note that the above expression is independent of the community of node i owing to the symmetry inherent in our model. Markov's inequality applied to the random variable $e^{-\lambda D_{ii}}$ then implies that for any $\epsilon > 0$,

$$\mathbb{P}(D_{ii} \leq \epsilon \log N) \leq e^{\lambda\epsilon\log N}\mathbb{E}e^{-\lambda D_{ii}} \leq N^{\lambda\epsilon-(1+o(1))\psi(s\lambda)}. \tag{11}$$

We note that $\psi(x)$ is a concave function with $\psi(0) = 0$ and

$$\psi'(0) = 2^{-(d-1)}\sum_m \sum_{r=0}^{d-1}\binom{d-1}{r}(d-1-2r)\alpha_{(r,d-r)}^{(m)} = 2^{-(d-1)}\xi,$$

where ξ is the assortativity defined by (5). Letting $s = \text{sgn}(\xi)$, it follows that

$$I := \sup_{x\in\mathbb{R}}\psi(x) = \sup_{\lambda\geq 0}\psi(s\lambda),$$

where I is the information quantity defined by (6). Given $d \geq 2$, we note that $-(d-1-2r)$ is positive for at least one $0 \leq r \leq d-1$, and the corresponding term in $\psi(x)$ decreases to $-\infty$ as x increases to ∞. On the other hand, $-(d-1-2r)$ is negative for at least one $0 \leq r \leq d-1$, and the corresponding term decreases to

$-\infty$ as x decreases to $-\infty$. Moreover, all of the terms are bounded from above. It follows that $\psi(x)$ attains its supremum on $\mathbb{R}$. If we assume that $I > 1$ and choose a small enough $\epsilon > 0$, then (11) implies that

$$\mathbb{P}(D_{ii} \leq \epsilon \log N) \leq N^{\lambda^* \epsilon - (1 + o(1))I} = o(N^{-1}),$$

where $\lambda^* = \arg\max_{\lambda \geq 0} \psi(s\lambda)$.

5.4 Assortativity

In this section, we provide an interpretation of assortativity in terms of the aggregate similarity matrix W in the asymptotic regime. This is stated in Proposition 1, which we prove below.

Proof (Proposition 1). First, we note that for $k \in \{-1, 1\}$ and $i \neq j$

$$w_{\text{in}} = \mathbb{E}[W_{ij} \mid \sigma_i = k, \sigma_j = k] = \sum_m \sum_{t \in \mathcal{T}_{d-2}} \binom{N/2 - 2}{t_k} \binom{N/2}{t_{-k}} p_{t + e_k + e_k}^{(m)},$$

where $\mathcal{T}_{d-2}$ is defined as in Sect. 5.3, and

$$w_{\text{out}} = \mathbb{E}[W_{ij} \mid \sigma_i = k, \sigma_j = -k] = \sum_m \sum_{t \in \mathcal{T}_{d-2}} \binom{N/2 - 1}{t_k} \binom{N/2 - 1}{t_{-k}} p_{t + e_k + e_{-k}}^{(m)}.$$

Applying the bounds $\left(1 - \frac{j}{n}\right)^j \frac{n^j}{j!} \leq \binom{n}{j} \leq \frac{n^j}{j!}$ and the scaling assumption (1),

$$w_{\text{in}} = \frac{\log N}{N} \cdot \frac{(1 + o(1))(d-1)}{2^{d-1}} \sum_m \sum_{t \in \mathcal{T}_{d-2}} \binom{d-2}{t_{-k}} \alpha_{t + e_k + e_k}^{(m)} = \Theta\left(\frac{\log N}{N}\right).$$

Similarly, $w_{\text{out}} = \Theta(\log N / N)$. By (7), for two communities of equal size we have

$$\mathbb{E}D_{ii} = \sum_{j \neq i : \sigma_i = \sigma_j} w_{\text{in}} - \sum_{j : \sigma_i \neq \sigma_j} w_{\text{out}} = \frac{N}{2}(w_{\text{in}} - w_{\text{out}}) - o(1). \tag{12}$$

Using (9) and (10), the expected value of D_{ii} can also be written as

$$\mathbb{E}D_{ii} = (1 + o(1))2^{-(d-1)} \xi \log N > 0$$

which combined with (12) implies the statement of the proposition.

5.5 Proof of Theorem 1

Lemma 5 shows that $D_{ii} \leq \epsilon \log N$ with probability $o(N^{-1})$. Taking union bound over i, we obtain $\min_{i \in [N]} D_{ii} \leq \epsilon \log N$ with probability $o(1)$. By Lemma 4, $\|W - \mathbb{E}W\|_2 \leq CM\sqrt{\log N}$ with probability $1 - O(N^{-11})$. Moreover, $w_{\text{in}} = \Theta(N^{-1} \log N)$ as shown in the proof of Proposition 1. By Lemmas 2 and 3, we then have $x^{\mathrm{T}} S x \geq \epsilon \log N - CM\sqrt{\log N} - N^{-1} \log N > 0$ with probability $o(1)$ for all $x \perp \sigma$ such that $\|x\|_2 = 1$. Application of Lemma 1 then concludes the proof.

6 Conclusions

In this work, we motivated and described the d-uniform multilayer inhomogeneous HSBM. We studied the problem of exact community recovery for the model using an SDP approach and the aggregate similarity matrix. For the symmetric case, our analysis provided a sufficient condition in terms of the information quantity I for community recovery. The generality of our model allows us to recover the sufficient conditions for some earlier models proposed in the literature.

Our treatment of the problem brings to the fore numerous related questions which are listed below:

- The assumption of symmetry on the parameters could be relaxed to make the hyperedge probabilities depend on the community labels. Additionally, it could be worthwhile to investigate asymmetry brought about by an imbalance in the community sizes.
- This work provides sufficient conditions for exact recovery based on the SDP approach. Necessary conditions for the multilayer HSBM model with the knowledge of the similarity matrix can be obtained using a methodology similar to [26] which will be addressed in a future publication.
- In this paper, the number of layers, M, is taken to be a constant independent on N. However, we expect that the analysis goes through when M grows slowly with N.
- The analysis of the SDP algorithm used here relies on the fact that there are just two communities. Extensions to a larger number of communities is a question worthy of investigation.

References

1. Abbé, E.: Community detection and stochastic block models: recent developments. J. Mach. Learn. Res. **18**, 1–86 (2018)
2. Ahn, K., Lee, K., Suh, C.: Hypergraph spectral clustering in the weighted stochastic block model. IEEE J. Sel. Top. Sign. Process. **12**(5), 959–974 (2018)
3. Alaluusua, K., Leskelä, L.: Consistent Bayesian community recovery in multilayer networks. In: IEEE International Symposium on Information Theory (ISIT), pp. 2726–2731 (2022)
4. Angelini, M.C., Caltagirone, F., Krzakala, F., Zdeborová, L.: Spectral detection on sparse hypergraphs. In: Annual Allerton Conference on Communication, Control, and Computing (2015)
5. Avrachenkov, K., Dreveton, M.: Statistical Analysis of Networks. Now Publishers Inc, Delft (2022)
6. Avrachenkov, K., Dreveton, M., Leskelä, L.: Community recovery in non-binary and temporal stochastic block models (2022). https://arxiv.org/abs/2008.04790
7. Bergman, E., Leskelä, L.: Connectivity of random hypergraphs with a given hyperedge size distribution (2022). https://arxiv.org/abs/2207.04799
8. Brusa, L., Matias, C.: Model-based clustering in simple hypergraphs through a stochastic blockmodel (2022). https://arxiv.org/abs/2210.05983

9. Chien, I., Lin, C.Y., Wang, I.H.: Community detection in hypergraphs: optimal statistical limit and efficient algorithms. In: International Conference on Artificial Intelligence and Statistics (AISTATS) (2018)

10. Chien, I.E., Lin, C.Y., Wang, I.H.: On the minimax misclassification ratio of hypergraph community detection. IEEE Trans. Inf. Theor. **65**(12), 8095–8118 (2019)

11. Cole, S., Zhu, Y.: Exact recovery in the hypergraph stochastic block model: a spectral algorithm. Linear Algebra Appl. **593**, 45–73 (2020)

12. Gaudio, J., Joshi, N.: Community detection in the hypergraph SBM: optimal recovery given the similarity matrix (2022). https://arxiv.org/abs/2208.12227

13. Ghoshdastidar, D., Dukkipati, A.: Consistency of spectral partitioning of uniform hypergraphs under planted partition model. In: Advances in Neural Information Processing Systems (NeurIPS) (2014)

14. Ghoshdastidar, D., Dukkipati, A.: A provable generalized tensor spectral method for uniform hypergraph partitioning. In: International Conference on Machine Learning (ICML) (2015)

15. Ghoshdastidar, D., Dukkipati, A.: Spectral clustering using multilinear SVD: analysis, approximations and applications. In: AAAI Conference on Artificial Intelligence (2015)

16. Ghoshdastidar, D., Dukkipati, A.: Consistency of spectral hypergraph partitioning under planted partition model. Ann. Stat. **45**(1), 289–315 (2017)

17. Gösgens, M.M., Tikhonov, A., Prokhorenkova, L.: Systematic analysis of cluster similarity indices: how to validate validation measures. In: International Conference on Machine Learning (ICML) (2021)

18. Guerrero-Sosa, J.D., Menéndez-Domínguez, V.H., Castellanos-Bolaños, M.E., Curi-Quintal, L.F.: Analysis of internal and external academic collaboration in an institution through graph theory. Vietnam J. Comput. Sci. **7**(04), 391–415 (2020)

19. Hajek, B., Wu, Y., Xu, J.: Achieving exact cluster recovery threshold via semidefinite programming. IEEE Trans. Inf. Theor. **62**(5), 2788–2797 (2016)

20. Hubert, L., Arabie, P.: Comparing partitions. J. Classif. **2**(1), 193–218 (1985)

21. Kamiński, B., Poulin, V., Prałat, P., Szufel, P., Théberge, F.: Clustering via hypergraph modularity. PLoS ONE **14**(11), 1–15 (2019)

22. Kamiński, B., Prałat, P., Théberge, F.: Community detection algorithm using hypergraph modularity. In: International Conference on Complex Networks and their Applications (2021)

23. Kamiński, B., Prałat, P., Théberge, F.: Hypergraph artificial benchmark for community detection (h-ABCD) (2022). https://arxiv.org/abs/2210.15009

24. Ke, Z.T., Shi, F., Xia, D.: Community detection for hypergraph networks via regularized tensor power iteration (2020). https://arxiv.org/abs/1909.06503

25. Kim, C., Bandeira, A.S., Goemans, M.X.: Community detection in hypergraphs, spiked tensor models, and sum-of-squares. In: International Conference on Sampling Theory and Applications (SampTA) (2017)

26. Kim, C., Bandeira, A.S., Goemans, M.X.: Stochastic block model for hypergraphs: statistical limits and a semidefinite programming approach (2018). https://arxiv. org/abs/1807.02884

27. Kivelä, M., Arenas, A., Barthelemy, M., Gleeson, J.P., Moreno, Y., Porter, M.A.: Multilayer networks. Journal of Complex Journal of Complex Journal of Complex Journal of Complex Journal of Complex J. Complex **2**(3), 203–271 (2014)

28. Kumar, T., Vaidyanathan, S., Ananthapadmanabhan, H., Parthasarathy, S., Ravindran, B.: A new measure of modularity in hypergraphs: theoretical insights and implications for effective clustering. In: International Conference on Complex Networks and their Applications (2019)

29. Kumar, T., Vaidyanathan, S., Ananthapadmanabhan, H., Parthasarathy, S., Ravindran, B.: Hypergraph clustering by iteratively reweighted modularity maximization. Appl. Netw. Sci. **5**(1), 1–22 (2020). https://doi.org/10.1007/s41109-020-00300-3

30. Lee, J., Kim, D., Chung, H.W.: Robust hypergraph clustering via convex relaxation of truncated MLE. IEEE J. Sel. Areas Inf. Theor. **1**(3), 613–631 (2020)

31. Lei, J., Chen, K., Lynch, B.: Consistent community detection in multi-layer network data. Biometrika **107**(1), 61–73 (2020)

32. Lesieur, T., Miolane, L., Lelarge, M., Krzakala, F., Zdeborová, L.: Statistical and computational phase transitions in spiked tensor estimation. In: 2017 IEEE International Symposium on Information Theory (ISIT), pp. 511–515. IEEE (2017)

33. Pal, S., Zhu, Y.: Community detection in the sparse hypergraph stochastic block model. Random Struct. Algorithms **59**, 407–463 (2021)

34. Pensky, M., Zhang, T.: Spectral clustering in the dynamic stochastic block model. Electr. J. Stat. **13**(1), 678–709 (2019)

35. Stephan, L., Zhu, Y.: Sparse random hypergraphs: Non-backtracking spectra and community detection (2022). https://arxiv.org/abs/2203.07346

36. Zhang, Q., Tan, V.Y.F.: Exact recovery in the general hypergraph stochastic block model. IEEE Trans. Inf. Theor. **69**(1), 453–471 (2023)

37. Zhen, Y., Wang, J.: Community detection in general hypergraph via graph embedding. J. Am. Stat. Assoc. 1–10 (2022)

38. Zhou, D., Huang, J., Schölkopf, B.: Learning with hypergraphs: clustering, classification, and embedding. In: Advances in Neural Information Processing Systems (NeurIPS) (2006)

The Myth of the Robust-Yet-Fragile Nature of Scale-Free Networks: An Empirical Analysis

Rouzbeh Hasheminezhad(✉), August Bøgh Rønberg, and Ulrik Brandes

ETH Zürich, Social Networks Lab, Zürich, Switzerland
{shashemi,ronberga,ubrandes}@ethz.ch

Abstract. In addition to their defining skewed degree distribution, the class of scale-free networks are generally described as robust-yet-fragile. This description suggests that, compared to random graphs of the same size, scale-free networks are more robust against random failures but more vulnerable to targeted attacks. Here, we report on experiments on a comprehensive collection of networks across different domains that assess the empirical prevalence of scale-free networks fitting this description. We find that robust-yet-fragile networks are a distinct minority, even among those networks that come closest to being classified as scale-free.

Keywords: robustness · scale-free networks · empirical data

1 Introduction

Scale-free networks are often portrayed as robust-yet-fragile, meaning that they are robust against random failures but more fragile against targeted attacks when compared to random graphs of the same size [6]. Although widely adopted, this characterization stems primarily from [1], where the authors base their conclusions on experiments with only two empirical networks and synthetic preferential-attachment networks generated using the model in [2].

In this paper, we investigate the empirical prevalence of robust-yet-fragile networks among those generally characterized as scale-free. To this end, we assemble an extensive collection of empirical networks from various domains, apply minimal preprocessing, and ensure maximal coverage of networks used in previous studies on the abundance of scale-free networks. We then utilize two state-of-the-art methods to identify the networks in our collection that are likely scale-free and evaluate their robustness compared to size-matching random graphs. Our main findings are as follows:

- The robust-yet-fragile networks in our collection are a small minority. Almost all of them come from the Route Views AS dataset, which captures BGP traffic at the autonomous systems level (AS level) on the Internet.

© The Author(s), under exclusive license to Springer Nature Switzerland AG 2023
M. Dewar et al. (Eds.): WAW 2023, LNCS 13894, pp. 99–111, 2023.
https://doi.org/10.1007/978-3-031-32296-9_7

– Although the networks from the Route Views AS dataset constitute the best potential candidates for classification as scale-free, the robust-yet-fragile networks are still no more than a distinct minority among them.

2 Data

To explore the robustness and scale-freeness of empirical networks, we use an extensive network collection fitting specific criteria, explained in this section. Most importantly, our collection contains, as much as possible, all the networks that have been considered previously in studies of scale-freeness.

2.1 Network Collection

We only consider networks represented as simple, undirected, and unweighted graphs. The reasons for this are two-fold: first, to avoid unprincipled decisions on handling additional levels of detail and complexity, and second, because our utilized robustness score is efficiently computable for such networks.

To ensure having a fair share of degree-heterogeneous and potentially scale-free networks, which are claimed to be prevalent empirically, we include every network used in [3,17,19] that matched the criterion described above. Moreover, we extended this combined collection with many other networks that met our criteria, based primarily on Netzschleuder [15], ICON [4], KONECT [11], and SNAP [13].

To increase the coverage of our dataset, we also include one-mode projections of two-mode networks if they are used as such in [3,17,19]. Finally, we include weighted networks under specific conditions, which we elaborate on later in this section.

2.2 Network Categorization

The following category labels are assigned to capture the contexts from which the networks within our collection originate: **technological**, **social**, **biological**, and **transportation**. In addition, we include a supplementary category, **auxiliary**, which contains only two word-adjacency networks that do not fit in the other categories. It is worth noting that each network in our collection exists in exactly one category.

2.3 Handling Weighted Networks

We include weighted networks only when the weights are non-negative integers corresponding to the multiplicities of the edges in a multigraph. In this case, we consider an unweighted version of the network, including all edges of non-zero weight, as this transformation does not affect the reachability and connectivity of the network. One such example from our collection is a network representing bus routes in India that describes which bus stops are connected by bus lines. In this network, the edge weights represent the number of overlapping bus lines that connect two bus stops. Here, we can safely ignore the edge weights because the network's reachability is indifferent if only one or twenty bus lines connect two stops.

2.4   Preprocessing

Before doing any analysis, we preprocess the networks in our collection. More specifically, for each network in our collection, all parallel edges and self-loops are removed, and anything not in the largest connected component is discarded. The last step is essential to ensure an equitable setting where the networks, being compared to each other in terms of robustness, are initially connected. The same preprocessing steps are applied before evaluating scale-freeness to ensure that the two evaluations assess the same networks, not slightly different ones. Finally, we discard from our collection all the networks that have fewer than a thousand vertices or edges after the aforementioned preprocessing steps have been applied, to make sure that the networks with which we work are large enough to make meaningful interpretations. This leaves us with more than nine hundred networks for performing our analysis.

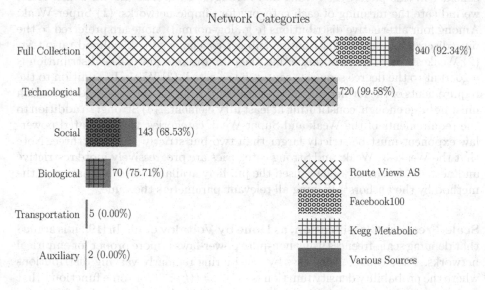

Fig. 1. The number of networks in each category (shown to the right of each bar) along with the proportion of the networks in each category that originate from one of the three largest network datasets within our collection (shown in the parentheses).

As shown in Fig. 1, an artifact of having a wide variety of sources for the network collection and including (as much as possible) all networks of [3,17,19] is that the networks in our collection are not evenly distributed among the different categories. Furthermore, approximately 92% of the networks in our collection originate from one of the Route Views AS [12], Facebook100 [18], and Kegg Metabolic datasets [8], which constitute the majority of the networks in the technological, social, and biological categories, respectively.

3   Scale-Freeness Analysis

Scale-free networks are those in which the degree sequence approximately follows a power-law distribution with a probability density function of $\Pr(k) = Ck^{-\alpha}$ where α is the power-law exponent and C is a constant [14]. There is a lot of controversy on how to classify networks regarding scale-freeness [7,9]; however, we use the following most representative methods from the literature.

3.1   Scale-Freeness Classification Methods

Scale-Freeness Classification, as Done by Broido et al. In [3], Broido et al. devised a classification of scale-freeness that, if not inconclusive, either declares a network not scale-free or puts it in one of five categories of scale-freeness, namely Super-weak, Weakest, Weak, Strong, and Strongest. For simple networks, as we consider in this paper, the Strong and Strongest categories are the same. Here, we indicate the meaning of each category for simple networks: (1) Super-Weak: Among four alternative distributions (e.g., log-normal), none are preferred to the pure discrete power-law distribution subject to Vuong's log-likelihood ratio test. (2) Weakest: The null hypothesis that the pure discrete power-law distribution is a good fit to the degree sequence cannot be rejected. (3) Weak: In addition to the requirements of the former category, the tail of the discrete power-law distribution must be large enough, constituting at least fifty elements. (4) Strong: In addition to the requirements of the Weak and Super-Weak categories, the estimated power-law exponent must be strictly larger than two but strictly less than three. Note that the Weakest, Weak, and Strong categories are progressively more restrictive and nested. Moreover, we have used the publicly available implementation of the method by the authors[1], keeping all relevant parameters the same.

Scale-Freeness Classification, as Done by Voitalov et al. In [19], it is argued that defining scale-freeness using non-pure power-laws is more proper for empirical networks. They capture this class by considering regularly varying distributions where the probability density function is $\Pr(k) = \ell(k)k^{-\alpha}$ for some function ℓ that varies slowly at infinity. Intuitively, due to the infinite degrees of freedom for choosing the function ℓ, hypothesis testing is impossible for this class of distributions, unlike the pure power-laws. However, the authors propose a scale-freeness classification based on the estimates of the power-law exponent α obtained by three consistent statistical estimators. For a given simple network, this method places the network in one of the following categories, which are non-overlapping and ordered from less restrictive to more restrictive: (1) Not power-law, (2) Hardly power-law, (3) Other power-law, and (4) Power-law with a divergent second moment. We use the publicly available implementation of this method by the authors[2], keeping all relevant parameters the same.

---

[1] https://github.com/adbroido/SFAnalysis.
[2] https://github.com/ivanvoitalov/tail-estimation.

3.2   Results

In Table 1, we present a summary of the main results when the two previously discussed scale-freeness classification methods generally agree on the extent to which the networks in our collection are scale-free. As can be seen in the table, both methods evaluate social and technological networks as the least and most scale-free networks, respectively. Additionally, Table 1 illustrates that most of the networks in our collection are placed within the highest scale-freeness classes according to both methods; however, this is largely due to the over-representation of technological networks, which predominantly represent the AS-level Internet architecture. When we consider the entire collection except for the technological networks, we observe that most networks are placed in the lowest categories of scale-freeness according to both methods.

Table 1. The columns indicate, respectively, the Weak (W), Super-Weak (SW), Not Scale-Free (NSF), and Inconclusive (INC) categories from the Broido et al. method. The rows indicate, respectively, the Divergent second moment Power-Law (DPL), Other Power-Law (OPL), Hardly Power-Law (HPL), and Not Power-Law (NPL) categories from the Voitalov et al. method. Each cell at the intersection of a row and a column indicates the percentage of networks at the intersection of the respective classes. Since we did not have any Strong, Super-Weak and Weak, or Weakest but not Weak networks in our collection, we utilized a more concise non-overlapping representation of scale-freeness classes, compared to the one originally defined by Broido et al.

	W	SW	NSF	INC	Total
DPL	71.38%	0.11%	7.23%	5.00%	83.72%
OPL	0.64%	0.00%	0.96%	0.64%	2.23%
HPL	0.32%	0.00%	4.47%	2.98%	7.77%
NPL	0.00%	0.00%	2.23%	4.04%	6.28%
Total	72.34%	0.11%	14.89%	12.66%	

(a) All

	W	SW	NSF	INC	Total
DPL	92.50%	0.00%	2.50%	4.72%	99.72%
OPL	0.00%	0.00%	0.00%	0.14%	0.14%
HPL	0.00%	0.00%	0.14%	0.00%	0.14%
NPL	0.00%	0.00%	0.00%	0.00%	0.00%
Total	92.50%	0.00%	2.64%	4.86%	

(b) Technological

	W	SW	NSF	INC	Total
DPL	2.27%	0.45%	22.73%	5.91%	31.36%
OPL	2.73%	0.00%	4.09%	2.27%	9.09%
HPL	1.36%	0.00%	18.64%	12.73%	32.73%
NPL	0.00%	0.00%	9.55%	17.27%	26.82%
Total	6.36%	0.45%	55.00%	38.18%	

(c) All without Technological

	W	SW	NSF	INC	Total
DPL	0.70%	0.70%	6.29%	2.10%	9.79%
OPL	1.40%	0.00%	3.50%	2.80%	7.69%
HPL	0.00%	0.00%	24.48%	18.18%	42.66%
NPL	0.00%	0.00%	13.99%	25.87%	39.86%
Total	2.10%	0.70%	48.25%	48.95%	

(d) Social

In the Appendix, we present the scale-freeness results for the categories where the evaluations of the Broido et al. and Voitalov et al. methods diverge from or disagree with one another. Additionally, we investigate whether the evaluation of these two methods on the social networks mostly being not scale-free persists when the Facebook100 networks, which constitute a noticeable fraction of the social networks, are not considered.

4   Robustness Analysis

4.1   Network Robustness

The robustness of a network is the invariance of one of its structural features, while the entities of the network, such as vertices and edges, are being removed subject to a given strategy [10]. In this paper, we focus on the largest connected component (or LCC for short) as the structural property of interest and consider only vertex removals by random failures and targeted attacks. During random failures, the vertices are removed uniformly at random. In targeted attacks, the vertices with the highest initial degree are removed first. To avoid potential bias, ties are broken randomly (e.g., when deciding which of the two vertices with the same degree should be removed first in a targeted attack, the one to be removed first is chosen randomly).

We quantify the robustness of an initially connected simple graph G by

$$R(G) = \frac{1}{n} \sum_{i=1}^{n} \frac{|\mathrm{LCC}\,(G\,[V\setminus\{v_1,\ldots,v_i\}])\,|}{|\mathrm{LCC}(G)|},$$

where $(v_1, v_2, \ldots, v_n)$ are the vertices of G in order of removal. Introduced in [16], this score captures not only the fraction of vertices in the largest connected component but also the rate at which this fraction shrinks during vertex removal.

4.2   Configuration

Here, we describe the experimental setup for an empirical network that has n vertices and m edges after the aforementioned preprocessing steps. Let $G(n, m)$ denote the uniform distribution over all networks with n vertices and m edges. We start from an empty baseline, and repeatedly sample from $G(n, m)$ until at least 96% of the vertices in the sampled network are in the largest connected component.[3] We then discard any vertices and edges not in the largest connected component, and add the network to the baseline. This process stops once 1000 networks have been added to the baseline. For each fixed vertex removal strategy considered in Sect. 4.1, we compute vertex removal orders on the empirical network and each of the baseline networks independently. Subsequently, for each network, the corresponding robustness scores are computed, given the respective vertex removal order. Finally, z-scores are used to compare the robustness of the empirical network to that of the baseline of size-matching random graphs under the given vertex removal strategy.[4] These z-scores are extensively used in the following subsections, where we present our results and their interpretation.

---

[3] We included in our collection only networks with enough density after preprocessing, such that this rejection-sampling step succeeds within at most one hundred attempts.

[4] The z-score, defined as $z = \sqrt{n}(x-\mu)/\sigma$, describes the relation of a value x to a group of n values that have mean μ and standard deviation σ. A positive or negative z-score captures, respectively, a tendency to obtain values above or below the reference mean. In light of this, we will use this score to evaluate how an empirical network's robustness compares to that of a baseline of size-matching random graphs, given a fixed vertex removal strategy.

4.3   Results

Here, we present in Fig. 2 the z-scores attained as described in the previous section. These z-scores reflect the robustness of the empirical networks in our collection compared to random graphs of the same size.

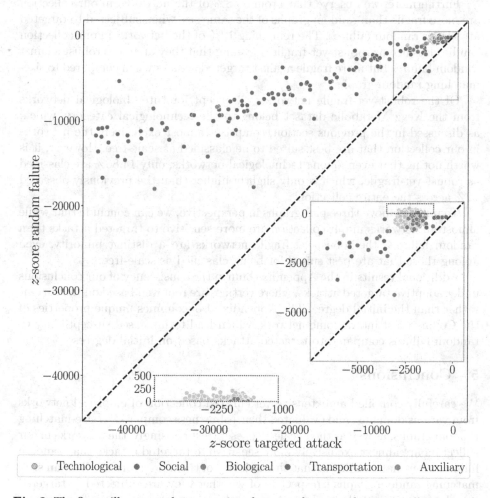

Fig. 2. The figure illustrates the comparison between the networks in our collection and size-matching random graphs in terms of their robustness against targeted attacks and random failures. Dark dashes depict the identity line. The first inset (marked with solid gray borders) zooms in on the area where most non-social networks are located. The second inset (marked with dark dotted borders) focuses on the region where networks are more robust against random failures but more vulnerable to targeted attacks (based on the initial degree), compared to the baseline of size-matching random graphs. Note that in favor of improving the clarity of the visualization, we have refrained from displaying the extreme outliers, which constitute 3% of all networks in our collection and attain (under both considered vertex removal strategies) highly negative z-scores. (Color figure online)

First, we notice that all the networks in our collection, except the Collins yeast interactome [5], are more sensitive to targeted attacks than random failures, as shown in Fig. 2 by the points above the identity line. We can also see that the networks in the social category of our collection constitute a dominant majority of the most fragile networks under both targeted attacks and random failures.

Furthermore, we observe that around 88% of the networks in our collection are more fragile than random graphs of the same size when subjected to targeted attacks or random failures. The remaining 12% of the networks in our collection can be classified as robust-yet-fragile, meaning that they are more robust against random failures but more fragile against targeted attacks when compared to size-matching random graphs.

Of the robust-yet-fragile networks, all except for three biological networks from the Kegg Metabolic dataset belong to the technological category, which, as discussed in the previous section, comprises almost exclusively the networks in our collection that are best suited to be classified as scale-free. However, it is worth noting that even among technological networks, only 14.86% are classified as robust-yet-fragile, which is only slightly higher than the previously observed 12% across the entire collection.

To put the above three paragraphs in perspective, we can conclude that while almost all networks in our collection are more sensitive to targeted attacks than random failures, the robust-yet-fragile networks are a distinct minority, even among those that are best suited for being classified as scale-free.

Additional results in the Appendix confirm the consistency of our conclusions under adaptive targeted attacks, where vertices are removed based on the current rather than the initial degree. The Appendix also examines unique properties of the Collins yeast interactome network, which lead to increased susceptibility to random failures compared to targeted attacks based on initial degrees.

5   Conclusions

We carefully compiled an extensive yet curated collection of empirical networks from various domains, and evaluated their robustness compared to size-matching random graphs, as well as their scale-freeness. Unsurprisingly, the networks in our collection are almost exclusively more sensitive to targeted attacks than random failures. Our main finding is that the majority of them are more fragile than size-matching random graphs irrespective of whether they are subjected to targeted attacks or random failures. In particular, this observation also applies to those networks that are the best candidates to be classified as scale-free. This is in stark contrast to the popular notion that a crucial feature of scale-free networks is their striking robustness against random failures. With few exceptions, neither engineered nor evolved networks in our collection compare favorably to random graphs regarding their robustness against random failures.

Availability of materials. Our code for collecting and preprocessing the network dataset, as well as the code for replicating our analysis and visualizations, are available at https://github.com/RouzbehHasheminezhad/WAW-2023-RYF.

6 Appendix

6.1 Scale-Freeness Classification: Further Analysis

In the following table, we present the scale-freeness classification results of the remaining categories, namely biological, transportation, and auxiliary. We can observe that the evaluations of the Broido et al. and Voitalov et al. methods are not necessarily aligned, for example, for the auxiliary and transportation categories. In fact, these evaluations sometimes disagree strongly, for instance, in the case of biological networks. The table also presents the scale-freeness classification results for the social category without considering the Facebook100 networks. It shows that even without these Facebook100 networks, 60% of all the social networks in our collection are evaluated as not scale-free, and around 47% of them are evaluated as hardly power-law or not power-law. This indicates that the earlier observed evaluation of both methods that scale-free networks do not constitute the strong majority of social networks is mostly consistent. We note that 99.98% of Facebook100 networks are hardly power-law or not power-law according to the Voitalov et al. method, while the evaluation of Broido et al. is inconclusive for 57.14% and places the rest in the not scale-free category (Table 2).

Table 2. The columns indicate, respectively, the Weak (W), Super-Weak (SW), Not Scale-Free (NSF), and Inconclusive (INC) categories from the Broido et al. method. The rows indicate, respectively, the Divergent second moment Power-Law (DPL), Other Power-Law (OPL), Hardly Power-Law (HPL), and Not Power-Law (NPL) categories from the Voitalov et al. method. Each cell at the intersection of a row and a column indicates the percentage of networks at the intersection of the respective classes. Since we did not have any Strong, Super-Weak and Weak, or Weakest but not Weak networks in our collection, we utilized a more concise non-overlapping representation of scale-freeness classes, compared to the one originally defined by Broido et al.

	W	SW	NSF	INC	Total
DPL	0.00%	0.00%	0.00%	0.00%	0.00%
OPL	40.00%	0.00%	0.00%	0.00%	40.00%
HPL	20.00%	0.00%	20.00%	20.00%	60.00%
NPL	0.00%	0.00%	0.00%	0.00%	0.00%
Total	60.00%	0.00%	20.00%	20.00%	

(a) Transportation

	W	SW	NSF	INC	Total
DPL	0.00%	0.00%	0.00%	0.00%	0.00%
OPL	0.00%	0.00%	50.00%	0.00%	50.00%
HPL	0.00%	0.00%	0.00%	50.00%	50.00%
NPL	0.00%	0.00%	0.00%	0.00%	0.00%
Total	0.00%	0.00%	50.00%	50.00%	

(b) Auxiliary

	W	SW	NSF	INC	Total
DPL	5.71%	0.00%	58.57%	14.29%	78.57%
OPL	2.86%	0.00%	4.29%	1.43%	8.57%
HPL	2.86%	0.00%	7.14%	0.00%	10.00%
NPL	0.00%	0.00%	1.43%	1.43%	2.86%
Total	11.43%	0.00%	71.43%	17.14%	

(c) Biological

	W	SW	NSF	INC	Total
DPL	2.22%	2.22%	20.00%	6.67%	31.11%
OPL	4.44%	0.00%	8.89%	8.89%	22.22%
HPL	0.00%	0.00%	22.22%	15.56%	37.78%
NPL	0.00%	0.00%	8.89%	0.00%	8.89%
Total	6.67%	2.22%	60.00%	31.11%	

(d) Social without Facebook100

6.2   Robustness: Further Analysis

To create Fig. 3, we used the same procedure as for Fig. 2, with the only variation being the usage of adaptive targeted attacks, in which vertices with a higher current degree are removed first rather than those with a higher initial degree.

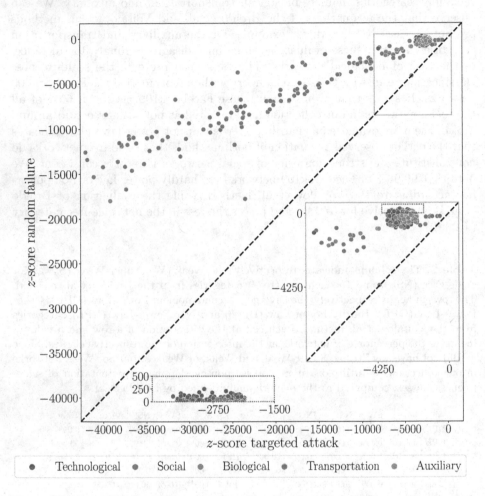

Fig. 3. The figure illustrates the comparison between the networks in our collection and size-matching random graphs in terms of their robustness against targeted attacks and random failures. Dark dashes depict the identity line. The first inset (marked with solid gray borders) zooms in on the area where most non-social networks are located. The second inset (marked with dark dotted borders) focuses on the region where networks are more robust against random failures but more vulnerable to targeted attacks (based on the current degree), compared to the baseline of size-matching random graphs. Note that in favor of improving the clarity of the visualization, we have refrained from displaying the extreme outliers, which constitute 3% of all networks in our collection and attain (under both considered vertex removal strategies) highly negative z-scores. (Color figure online)

As shown in Fig. 3, all networks, including the Collins yeast interactome network, exhibit increased sensitivity to adaptive targeted attacks compared to random failures. This is evident from the fact that the markers for the networks across different categories are above the identity line, and the marker for the Collins yeast interactome network is very slightly above the identity line. We also observe that the networks from the social category continue to constitute a noticeable fraction of the most fragile networks in the entire collection.

Our analysis shows that 88% of the networks in our collection are more fragile than size-matching random graphs, while the remaining 12% are robust-yet-fragile. This is consistent with our previous results in Sect. 4, with all robust-yet-fragile networks except for three Kegg Metabolic networks being from the technological category, which comprises almost exclusively the networks in our collection best suited for classification as scale-free. The percentage of robust-yet-fragile networks among technological networks is 14.86%, similar to the findings of Sect. 4.

Based on the above two paragraphs, we can conclude that all the networks in our collection are more sensitive to adaptive targeted attacks than they are to random failures. Furthermore, those networks that can be characterized as robust-yet-fragile constitute only a distinct minority, even among those networks that are the best candidates for being classified as scale-free, which is in line with our conclusions in Sect. 4.

6.3   The Curious Case of Collins Yeast Interactome

A protein complex is a group of proteins that work together to perform a specific biological function. In the Collins yeast interactome network, individual proteins in budding yeast are represented as vertices, and edges connect all proteins that are part of the same protein complex to each other [5]. Thus, the structure of this network is characterized by the presence of dense, interconnected clusters, with larger clusters comprising high-degree vertices of proteins that interact mostly with other proteins within the same complex. These clusters are interconnected mainly through connectivity hubs, which are proteins that do not necessarily interact with many proteins within their protein complex, but with potentially few proteins from different complexes. In Fig. 4 on the next page, we present a visualization of the Collins yeast interactome network, in which the discussed network structure is apparent.

This structure has important implications for the robustness of the network. To observe this, note that high-degree vertices are found within dense clusters of similarly high-degree vertices, so targeted attacks which remove high-degree vertices first would mainly focus on disintegrating these high-density clusters, while leaving the low-degree vertices that interconnect them intact. In contrast, random failures are more likely to hit these low-degree interconnecting vertices, resulting in a faster disintegration of the network into many smaller, disjoint clusters. This explains the higher susceptibility of the robustness of the Collins yeast interactome network to random failures compared to targeted attacks, as opposed to any other network in our collection.

Fig. 4. The largest connected component of the Collins yeast interactome network. The size of each vertex is proportional to its degree in the network.

References

1. Albert, R., Jeong, H., Barabási, A.L.: Error and attack tolerance of complex networks. Nature **406**(6794), 378–382 (2000)
2. Barabási, A.L., Albert, R.: Emergence of scaling in random networks. Science **286**(5439), 509–512 (1999)
3. Broido, A.D., Clauset, A.: Scale-free networks are rare. Nat. Commun. **10**(1), 1017 (2019)
4. Clauset, A., Tucker, E., Sainz, M.: The Colorado index of complex networks (2016). https://icon.colorado.edu/
5. Collins, S.R., et al.: Toward a comprehensive atlas of the physical interactome of saccharomyces cerevisiae. Mol. Cell. Proteomics **6**(3), 439–450 (2007)
6. Doyle, J.C., et al.: The "robust yet fragile" nature of the Internet. Proc. Natl. Acad. Sci. **102**(41), 14497–14502 (2005)
7. Holme, P.: Rare and everywhere: perspectives on scale-free networks. Nat. Commun. **10**(1), 1016 (2019)

8. Huss, M., Holme, P.: Currency and commodity metabolites: their identification and relation to the modularity of metabolic networks. IET Syst. Biol. **1**(5), 280–285 (2007)
9. Jacomy, M.: Epistemic clashes in network science: mapping the tensions between idiographic and nomothetic subcultures. Big Data Soc. **7**(2) (2020)
10. Klau, G.W., Weiskircher, R.: Robustness and resilience. In: Brandes, U., Erlebach, T. (eds.) Network Analysis. LNCS, vol. 3418, pp. 417–437. Springer, Heidelberg (2005). https://doi.org/10.1007/978-3-540-31955-9_15
11. Kunegis, J.: KONECT: the Koblenz network collection. In: Proceedings of the 22nd International Conference on World Wide Web, WWW 2013 Companion, pp. 1343–1350. Association for Computing Machinery (2013)
12. Leskovec, J., Kleinberg, J., Faloutsos, C.: Graphs over time: densification laws, shrinking diameters and possible explanations. In: Proceedings of the Eleventh ACM SIGKDD International Conference on Knowledge Discovery in Data Mining, KDD 2005, pp. 177–187. Association for Computing Machinery (2005)
13. Leskovec, J., Krevl, A.: SNAP Datasets: Stanford large network dataset collection (2014). http://snap.stanford.edu/data
14. Li, L., Alderson, D., Doyle, J.C., Willinger, W.: Towards a theory of scale-free graphs: definition, properties, and implications. Internet Math. **2**(4), 431–523 (2005)
15. Peixoto, T.P.: The Netzschleuder network catalogue and repository (2020). https://networks.skewed.de/
16. Schneider, C.M., Moreira, A.A., Andrade, J.S., Havlin, S., Herrmann, H.J.: Mitigation of malicious attacks on networks. Proc. Natl. Acad. Sci. **108**(10), 3838–3841 (2011)
17. Serafino, M., et al.: True scale-free networks hidden by finite size effects. Proc. Natl. Acad. Sci. **118**(2), e2013825118 (2021)
18. Traud, A.L., Mucha, P.J., Porter, M.A.: Social structure of Facebook networks. Phys. A: Stat. Mech. Appl. **391**(16), 4165–4180 (2012)
19. Voitalov, I., van der Hoorn, P., van der Hofstad, R., Krioukov, D.: Scale-free networks well done. Phys. Rev. Res. **1**(3), 033034 (2019)

A Random Graph Model for Clustering Graphs

Fan Chung and Nicholas Sieger$^{(\boxtimes)}$

University of California, San Diego, La Jolla, CA 92037, USA
{fan,nsieger}@ucsd.edu

Abstract. We introduce a random graph model for clustering graphs with a given degree sequence. Unlike previous random graph models, we incorporate clustering effects into the model without any geometric conditions. We show that random clustering graphs can yield graphs with a power-law expected degree sequence, small diameter, and any desired clustering coefficient. Our results follow from a general theorem on subgraph counts which may be of independent interest.

Keywords: Random graphs · clustering coefficient · scale-free networks · Chung-Lu model

1 Introduction

A wide range of real-world networks possess a *small world phenomenon* where every node is within a short distance of every other node, and friends of friends are themselves friends. In addition, the degree sequence of such networks follows a power-law distribution [16]. Most random graph models have a degree sequence with power-law distribution with small diameter [5], but do not exhibit clustering behavior. One way to introduce clustering in a random graph model is to impose geometric constraints by embedding a graph into a particular geometric space such as a high-dimensional manifold [1,7,8,14] or hyperbolic surface [15]. However, the natural graph distances of many information networks, social networks, or biological networks [4,5,13,16,17] are inconsistent with these extrinsic geometric constraints.

The goal of this paper is to introduce a general random graph model without any additional geometric constraints so that the only geometry is that which is naturally present in the graph itself. In fact, we prove a stronger result. Our models works for arbitrary degree sequences, not just power-law degree sequences, and we analyze the subgraph counts of an arbitrary fixed graph in our model, rather than merely the graphs needed to define the clustering coefficient. As a consequence, our model can generate graphs with power-law degree sequence, small diameter, and any desired constant clustering coefficient.

Our work begins with the *Chung-Lu* model, denoted by $\mathcal{G}(\mathbf{d})$, which was introduced in [3]. The construction of $\mathcal{G}(\mathbf{d})$ begins with a vector $\mathbf{d} = (\mathbf{d}_1, \ldots, \mathbf{d}_n)$

M. Dewar et al. (Eds.): WAW 2023, LNCS 13894, pp. 112–126, 2023.
https://doi.org/10.1007/978-3-031-32296-9_8

and adds an edge between vertices u and v independently with probability proportional to $\mathbf{d}_u \mathbf{d}_v$. Under some very weak conditions on the vector $\mathbf{d}$, $\mathcal{G}(\mathbf{d})$ is a distribution on graphs such that the expected degree sequence is $\mathbf{d}$. Furthermore, $\mathcal{G}(\mathbf{d})$ has a long list of desirable properties similar to the Erdős-Rényi model [12], from the existence of a giant component [10], to small average distance and diameter [9], and explicit bounds on its spectrum [11]. We note that the Chung-Lu model requires the maximum degree in $\mathbf{d}$ to be bounded, specifically that maximum degree is at most the square root of the volume (the sum of the degrees). We will assume that this condition holds for the remainder for the paper.

Our work builds on the Chung-Lu model by adding an additional step to the construction. Given a vector $\mathbf{d}$ and an additional parameter $\gamma \in (0,1)$, we construct a graph as follows:

1. Generate a graph G_0 drawn from the Chung-Lu model with degree vector $\mathbf{d}$.
2. For each path uv, vw in G_0, add the edge uw independently with probability γ.

The resulting graph G is said to be a *random clustering graph* drawn from the distribution $\mathcal{C}(\gamma, \mathbf{d})$. The purpose of this paper is to analyze $\mathcal{C}(\gamma, \mathbf{d})$ and in particular, analyze its clustering coefficient. The key obstacle in analyzing the clustering coefficient is the dependencies between the edges in $\mathcal{C}(\gamma, \mathbf{d})$. We characterize a class of graphs called extension configurations which allow us to reduce the problem of subgraph counts in $\mathcal{C}(\gamma, \mathbf{d})$ to subgraph counts in $\mathcal{G}(\mathbf{d})$, bypassing the dependencies between edges. Via these extension configurations, we can compute the expected subgraph count of a given graph in $\mathcal{C}(\gamma, \mathbf{d})$ and then show that subgraph counts are concentrated around the mean.

The paper is organized as follows. We give notation and some basic definitions in Sect. 2. We formally define the Chung-Lu model and analyze its homomorphism counts in Sect. 3. We then define our random clustering graphs model $\mathcal{C}(\gamma, \mathbf{d})$ and present a few basic properties in Sect. 4. In Sect. 5, we prove our main result, Theorem 4, on the concentration of subgraph counts, and then apply Theorem 4 to find the clustering coefficient of $\mathcal{C}(\gamma, \mathbf{d})$ and show concentration.

2   Preliminaries

All graphs will be finite and simple unless indicated otherwise. We will write $V(G)$ for the vertex set of a graph G and $v(G) = |V(G)|$ for number of vertices. We also write $E(G)$ for the edge set of a graph G and $e(G) = |E(G)|$ of the number of edges. We write $N_G(v)$ for the neighborhood of a vertex $v \in V(G)$, and $\deg_G(v)$ for the degree of $v \in V(G)$.

A graph H is a *subgraph* of a graph G if there is an injective map $\phi : H \hookrightarrow G$ such that

$$uv \in E(H) \implies \phi(u)\phi(v) \in E(G).$$

We will refer to the map ϕ as a *graph homomorphism*. The number of graph homomorphisms from H to G will be denoted by $\hom(H, G)$. For a specific injection $\psi : V(H) \hookrightarrow V(G)$, we write $H \curvearrowright_\phi G$ for the proposition that ϕ is a graph homomorphism of H to G.

The *clustering coefficient* of a graph G is

$$C_1(G) = 3\frac{\hom(K_3, G)}{\hom(P_2, G)}$$

where P_2 denotes the path with 2 edges.

For a vector $\mathbf{d}$ with n entries, the *kth-order expected volume* of $\mathbf{d}$, denoted $\mathrm{Vol}_k(\mathbf{d})$, is

$$\sum_{i=1}^{n} (\mathbf{d}_i)^k$$

and we write $\mathrm{Vol}(\mathbf{d})$ to denote the 1st order volume. The *second-order average degree* of a vector $\mathbf{d}$, denoted by $\tilde{d}(\mathbf{d})$, is $\mathrm{Vol}_2(\mathbf{d})/\mathrm{Vol}(\mathbf{d})$.

Finally, we write $\mathbf{1}_E$ for the indicator random variable of the event E. We write $f = o(g)$ for two non-negative functions $f : \mathbb{N} \to \mathbb{R}$ and $g : \mathbb{N} \to \mathbb{R}$ if $\lim_{n\to\infty} \frac{f(n)}{g(n)} = 0$. We write $f = O(g)$ if there is a constant C independent of n such that $f(n) \leq Cg(n)$ for all $n \geq 1$, we write $f = \Omega(g)$ if there is a constant C such that $f(n) \geq Cg(n)$ for all $n \geq 1$, and write $f = \Theta(g)$ if $f = O(g)$ and $f = \Omega(g)$. Finally, we say that an event E holds *with high probability* if $\mathbb{P}[E] = 1 - o(1)$.

If $\mathcal{D}$ is a distribution on graphs we write $D \hookleftarrow \mathcal{D}$ to indicate that the graph D is drawn from the distribution $\mathcal{D}$.

3   Homomorphism Counts in the Chung-Lu Model

We now formally define the Chung-Lu model and analyze homomorphism counts in the Chung-Lu model.

Definition 1. *[3] Let $\mathbf{d}$ be a vector of length n. The Chung-Lu random graph, denoted $\mathcal{G}(\mathbf{d})$, is the distribution on graphs on the vertex set $[n]$ where each edge $u \in [n]$ is adjacent to $v \in [n]$ independently with probability*

$$\frac{\mathbf{d}_u \mathbf{d}_v}{\mathrm{Vol}(\mathbf{d})}.$$

Rather than control the entries of the vector $\mathbf{d}$ directly, we instead impose conditions on the expected volumes $\mathrm{Vol}_k(\mathbf{d})$.

Definition 2. *A vector $\mathbf{d}$ satisfies the (a, b, Δ)-volume growth condition if there are constants $a \in [0, \frac{1}{2})$ and $b \in [0, 1 - a]$ such that $\mathrm{Vol}_k(\mathbf{d}) = \Theta(\mathrm{Vol}(\mathbf{d})^{ak+b})$ for every k such that $0 < k \leq \Delta$.*

Intuitively, the (a, b, Δ)-Volume growth condition says that the moments $\mathrm{Vol}_k(\mathbf{d})$ do not grow too quickly but do grow, at least for small values of k. The (a, b, Δ)-volume growth condition arises from our proofs, particularly Theorem 3. There are several examples of vectors $\mathbf{d}$ which satisfy the (a, b, Δ)-Volume Growth condition for various values of a and b.

Example 1. Let $\mathbf{d}$ be a constant vector with entries n^{1-c} for some fixed value $c \in (0, 1)$. Then,

$$\text{Vol}_k(\mathbf{d}) = n^{1+(1-c)k}$$

for each k such that $k \geq 1$. Notice that $\text{Vol}(\mathbf{d}) = n^{2-c}$. Therefore,

$$\text{Vol}_k(\mathbf{d}) = n^{1+(1-c)k} = n^{(2-c)\left(\frac{1}{2-c} + \frac{1-c}{2-c}k\right)} = \text{Vol}(\mathbf{d})^{\frac{1-c}{2-c}k + \frac{1}{2-c}}.$$

Since $\frac{1-c}{2-c} + \frac{1}{2-c} = 1$, and $\frac{1-c}{2-c} < \frac{1}{2}$ for $c > 0$, it follows that $\mathbf{d}$ satisfies the $\left(\frac{1-c}{2-c}, \frac{1}{2-c}, \infty\right)$ volume growth condition. For such a choice of vector $\mathbf{d}$, the model $\mathcal{G}(\mathbf{d})$ corresponds to the Erdős-Rényi random graph $G(n, n^{-c})$.

Example 2. Fix a large $\alpha \in \mathbb{R}_{>0}$ and $\beta \in (2, 3)$. For a vector $\mathbf{d}$, let $y_{\mathbf{d}}(x)$ denote the number of indices i such that $\mathbf{d}_i = x$. Let $\mathbf{d}$ be a vector such that $\ln(y_{\mathbf{d}}(x)) = \alpha - \beta \ln(x)$, i.e. $\mathbf{d}$ satisfies a power-law with exponent $2 < \beta < 3$. By results in [2],

$$\text{Vol}(\mathbf{d}) = \Theta(e^{\alpha})$$

and for $k > 1$,

$$\text{Vol}_k(\mathbf{d}) = \Theta\left(e^{\left(\frac{\alpha(k+1)}{\beta}\right)}\right).$$

Hence,

$$\text{Vol}_k(\mathbf{d}) = \Theta\left(e^{\left(\alpha\left(\frac{k}{\beta} + \frac{1}{\beta}\right)\right)}\right) = \Theta\left(\text{Vol}(\mathbf{d})^{\frac{k}{\beta} + \frac{1}{\beta}}\right)$$

Since $\frac{1}{\beta} + \frac{1}{\beta} \leq 1$ and $\frac{1}{\beta} < \frac{1}{2}$ for $\beta > 2$, it follows that $\mathbf{d}$ satisfies the $(1/\beta, 1/\beta, \infty)$ volume growth condition. In this case, the model $\mathcal{G}(\mathbf{d})$ is a Chung-Lu random graph with a power-law degree distribution.

With this definition in hand, we can show that the expected number of homomorphisms of H into $\mathcal{G}(\mathbf{d})$ is determined asymptotically by the degree sequence of H and the expected volumes of $\mathbf{d}$. The proof can be found in the full version of the paper.

Lemma 1. *Let H be a fixed graph. If $\mathbf{d} \in \mathbb{R}^n_{>0}$ is a vector which satisfies the $(a, b, \Delta(H))$-volume growth condition with $b > 0$, then the subgraph count of H in $\mathcal{G}(\mathbf{d})$ is asymptotically equal to the following expression:*

$$(1 - o(1)) \frac{\prod_{v \in V(H)} \text{Vol}_{\deg_H(v)}(\mathbf{d})}{\text{Vol}(\mathbf{d})^{e(H)}} \leq \mathbb{E}\left[\text{hom}(H, G)\right] \leq \frac{\prod_{v \in V(H)} \text{Vol}_{\deg_H(v)}(\mathbf{d})}{\text{Vol}(\mathbf{d})^{e(H)}}$$

The above lemma will serve as our primary estimate for subgraph counts of H in $\mathcal{G}(\mathbf{d})$ for the remainder of the paper.

4   Random Clustering Graph Model

Here we define our random clustering graphs and present several basic properties.

Definition 3. *Let* $\mathbf{d} \in \mathbb{R}_{\geq 0}^n$ *be a vector and fix* $\gamma \in [0,1]$. *A random clustering graph, whose distribution we denote by* $\mathcal{C}(\gamma, \mathbf{d})$, *is a graph constructed via the following procedure:*

1. *Draw a graph* $G_0 \hookleftarrow \mathcal{G}(\mathbf{d})$
2. *If* $(u, v) \notin E(G_0)$, *for each vertex* $w \in V(G_0)$ *such that* (u, w) *and* $(w, v) \in E(G_0)$, *flip an independent coin of bias* γ *to add the edge* (u, v).

When working with a graph $G \hookleftarrow \mathcal{C}(\gamma, \mathbf{d})$, we henceforth write G_0 for the graph generated in the first stage and G for the graph found in the second stage.

Given the close relationship of $\mathcal{C}(\mathbf{d}, \gamma)$ to $\mathcal{G}(\mathbf{d})$, we first analyze the expected degree sequence. The proof can be found in the full version of the paper.

Theorem 1. *If* $\mathbf{d}$ *is a vector satisfying* $\frac{\mathbf{d}_{\max}^2 \operatorname{Vol}_2(\mathbf{d})}{\operatorname{Vol}(\mathbf{d})^2} = o(1)$, *then the degree of a vertex* v *in a random clustering graph is* $(1 - o(1)) \left(1 + \gamma \tilde{d}(\mathbf{d})\right) \mathbf{d}_v$ *in expectation.*

The condition $\frac{\mathbf{d}_{\max}^2 \operatorname{Vol}_2(\mathbf{d})}{\operatorname{Vol}(\mathbf{d})^2} = o(1)$ can be thought of as stating that all probabilities in $\mathcal{C}(\gamma, \mathbf{d})$ are $o(1)$.

Theorem 1 tells us that, in expectation, the degree sequence of $\mathcal{C}(\gamma, \mathbf{d})$ is a scaled version of $\mathbf{d}$. Thus the degree sequence has the same overall shape as $\mathbf{d}$, i.e. if $\mathbf{d}$ is constant, so is the degree sequence of $\mathcal{C}(\gamma, \mathbf{d})$ in expectation.

As an immediate consequence, we can show that a wide range of degree sequences are realizable in expectation. The proof can be found in the full version of the paper.

Theorem 2. *Let* $\mathbf{d} \in \mathbb{R}_{\geq 0}^n$ *be a fixed vector such that* $\frac{\mathbf{d}_{\max}^2 \operatorname{Vol}_2(\mathbf{d})}{\operatorname{Vol}(\mathbf{d})^2} = o(1)$. *If* $\gamma \in (0, 1)$ *is fixed, then there is a constant* $c \in (0, 1)$ *such that a random clustering graph* $G \hookleftarrow \mathcal{C}(\gamma, c\mathbf{d})$ *has expected degree sequence given by* $\mathbf{d}$. *Specifically, we can take* $c = \frac{\sqrt{1 + 4\gamma \tilde{d}(\mathbf{d})} - 1}{2\gamma \tilde{d}(\mathbf{d})}$.

Thus we can simulate any slowly growing degree sequence in expectation, just as in $\mathcal{G}(\mathbf{d})$.

5   Homomorphism Counts

In this section, we define a class of graphs which control homomorphism counts in $\mathcal{C}(\gamma, \mathbf{d})$ and then give the expected number of homomorphisms of a fixed graph in Theorem 3 and then prove concentration of homomorphism counts in $\mathcal{C}(\gamma, \mathbf{d})$ in Theorem 4.

5.1   Extension Configurations

Here we define the graphs needed to describe homomorphism counts in $\mathcal{C}(\gamma, \mathbf{d})$. The precise relationship is found in Lemma 2.

To gain intuition for how to compute homomorphism counts in $\mathcal{C}(\gamma, \mathbf{d})$, consider the example of the triangle K_3 with a fixed injection $\phi : V(K_3) \to [n]$.

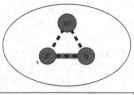 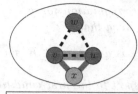 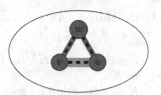

Case 1: uv is already in G_0

Case 2: there is a vertex $x \notin \phi(K_3)$ adjacent to u and v and the coin flip comes up heads

Case 3: there is a vertex $w \in \phi(K_3)$ adjacent to u and v and the coin flip comes up heads.

Fig. 1. The three cases for an edge $uv \in K_3$ to exist in G. Edges in K_3 are drawn in black dashed lines, edges in G_0 are drawn in solid red lines, and edges added into G are in thick green lines. (Color figure online)

For each edge $uv \in E(K_3)$, Fig. 1 illustrates the three ways in which an edge $uv \in K_3$ can appear in G.

More formally, for $\phi : V(H) \hookrightarrow [n]$ to induce a copy of H as a subgraph in G, at least one of the following must hold for every $uv \in H$.

1. $\phi(u)\phi(v) \in E(G_0)$.
2. There is a vertex $w \in V(H)$ such that the edges $\phi(u)\phi(w)$ and $\phi(v)\phi(w)$ appear in G_0.
3. There is a vertex x not in the image of H under ϕ such that the edges $\phi(u)x$ and $\phi(v)x$ appear in G_0.

Now it is entirely possible that a single edge $uv \in H$ can satisfy more than one of the above conditions. In the subgraph appearing in Fig. 2, several edges of K_3 can be added twice. Thus, we wish to consider a "minimal subgraph" which contains all the edges needed to form a copy of our desired subgraph H and no more edges than needed. To that end, we have the following key definition:

Fig. 2. A "redundant" extension configuration. Red solid edges indicate the edges in G_0 and green edges are those added by coin flips. The edge uv is added twice as it appears in G_0 and is added by the path uxv. The edge vw is added twice, once by the path vuw and once by the path vyw. (Color figure online)

Definition 4. *An extension configuration for a graph H is a graph K on the vertex set $V(H) \sqcup \{v_1, \ldots, v_l\}$ such that*

- *For every $uv \in H \setminus K$, there is a vertex $x \in K$ such that the edges ux, vw are in K.*
- *K is edge-minimal with respect to the first condition.*

The *external vertices* of an extension configuration K of H, denoted $Ex(K)$, are the vertices $V(K) \setminus V(H)$. Let $ex(K) = |Ex(K)|$. Let $Ext(H)$ denote the set of extension configurations of a graph H.

Considering the triangle again, we can exactly enumerate the extension configurations for K_3 (up to relabeling the vertices) as shown in Fig. 3.

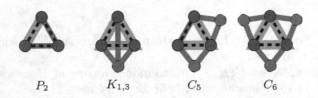

P_2 $K_{1,3}$ C_5 C_6

Fig. 3. The 4 extension configurations for K_3 up to relabeling of vertices, i.e. the elements of $Ext(K_3)$. The edges of K_3 are labeled by dashed black lines and the edges of the extension configuration are in solid red. Each of the dashed lines is either in the extension configuration, has a unique external vertex adjacent to both endpoints, or has at least one internal vertex adjacent to both endpoints. (Color figure online)

We can observe that the "redundant" subgraph in Fig. 2 is actually the union of the two extension configurations P_2 and C_5. Note that extension configurations can include edges on the same vertex set as the desired graph which are not part of the original graph. Figure 4 shows an example:

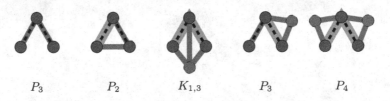

P_3 P_2 $K_{1,3}$ P_3 P_4

Fig. 4. The extension configurations for P_2. Observe in the second example that extension configurations of H can include edges on the same vertex set of H which are not edges in H itself.

The one aspect of the construction of $G \sim \mathcal{C}(\gamma, \boldsymbol{d})$ that we have yet to consider is the role of the coin flips. We now integrate this second stage of randomness:

Definition 5. *For a fixed extension configuration K with $ex(K) = l$ and an injective graph homomorphism $\psi : V(K) \hookrightarrow [n]$, let $B_{K,\psi}$ denote the event that the following occurs:*

1. *For each edge $uv \in E(K)$, the edge $\psi(u)\psi(v)$ is in $E(G_0)$.*
2. *For each $i \in [l]$ and each $uv \in N_K(v_i) \cap E(H)$, the coin flip for the path $\psi(u)\psi(p_i)\psi(v)$ comes up heads and the coin flips for any vertex $w \in V(H)$ such that uwv is a path in K come up tails.*
3. *If $uv \in E(H) \setminus E(K)$ and there is no $v_i \in Ek(K)$ such that $N_K(v_i)$ which contains both u and v, then at least one of coin flips for vertices $w \in V(H)$ such that uwv is a path in K comes up heads.*

Remark 1. The definition of an extension configuration does not distinguish between vertices $x \in \phi(H)$ and vertices $x \notin \phi(H)$, yet the event $B_{K,\psi}$ does make this distinction. As a consequence of edge-minimality of an extension configuration, for any edge $uv \in H$, there can be at most one external vertex $x \in Ex(K)$ such that $ux, vx \in K$. However, there can be two (or more) vertices $x, y \in H$ such that $ux, vx, uy, vy \in K$ and yet the extension configuration remains edge-minimal. See Fig. 5 for an example. Thus, edges going to external vertices and internal edges need to be treated differently.

Fig. 5. An extension configuration which has an edge added twice. Notice that no edge can be removed.

With this definition in hand, we can show that extension configurations exactly capture homomorphisms of graphs into $\mathcal{C}(d, \gamma)$:

Lemma 2. *Fix a graph H and an injection $\phi : H \hookrightarrow [n]$. If G is a random clustering graph on the vertex set $[n]$, then the probability that ϕ is a graph homomorphism of H to G is equal to the probability that at least one of the events $B_{K,\psi}$ occurs for some extension configuration $K \in Ext(H)$ and $\psi : K \hookrightarrow [n]$ which extends ϕ. Formally, we have*

$$\mathbb{P}[H \curvearrowright_\phi G] = \mathbb{P}\left[\bigvee_{K \in Ext(H)} \bigvee_{\substack{\psi : V(K) \to [n] \\ \psi|_{V(H)} = \phi}} B_{K,\psi} \right]$$

Proof. By definition, if any of the events $B_{K,\psi}$ occur, then $H \curvearrowright_\phi G$ must also occur. Hence,

$$\mathbb{P}\left[H \curvearrowright_\phi G\right] \geq \mathbb{P}\left[\bigvee_{K \in Ext(H)} \bigvee_{\substack{\psi:V(K)\to[n] \\ \psi|_{V(H)}=\phi}} B_{K,\psi}\right].$$

Now if $H \curvearrowright_\phi G$, then one of the following must be true for each edge $uv \in E(H)$:

- $\phi(u)\phi(v) \in E(G_0)$
- There is a vertex $w \in [n]$ such that $\phi(u)\phi(w), \phi(v)\phi(w) \in E(G_0)$ and the coin flip for the corresponding path comes up heads.

We define a graph K via the following procedure. For each edge $uv \in E(H)$:

- If $\phi(u)\phi(v) \in E(G_0)$, add uv to K.
- If there is a vertex $x \in [n]$ such that $\phi(u)x, \phi(v)x \in E(G_0)$ and the coin flip for the path $\phi(u)x\phi(v)$ comes up heads, choose such a vertex x arbitrarily. If x has not already been chosen previously in the procedure, add a new vertex w to K. If x has already been chosen, set w to be the vertex created when x was first chosen. Then add the edges uw, vw to K.

The external vertices of K are defined to be the new vertices w added in the second step.

We claim that K contains an extension configuration for H. Indeed, if $uv \in E(H) \setminus E(K)$ is not in the neighborhood of any external vertex in K, then there is no vertex $x \in [n] \setminus \phi(V(H))$ such that $\phi(u)x, \phi(v)x \in E(G_0)$ and the coin flip came up heads. Thus, there must be a vertex $w \in V(H)$ such that $\phi(v)\phi(w), \phi(u)\phi(w) \in E(G_0)$. As we include all such edges, it follows that K satisfies the first condition of an extension configuration. We can then may remove any redundant edges $uv \in K$ to produce a genuine extension configuration K'.

We observe that one can choose vertices $x \in [n]$ so that any extension configuration can arise in the above procedure. Finally, one can easily define an injection $\psi : V(K') \hookrightarrow [n]$ such that $\psi|_{V(H)} = \phi$ by following the choices of x and w in the second step. Therefore, we can conclude that

$$\mathbb{P}\left[H \curvearrowright_\phi G\right] \leq \mathbb{P}\left[\bigvee_{K \in Ext(H)} \bigvee_{\substack{\psi:V(K)\to[n] \\ \psi|_{V(H)}=\phi}} B_{K,\psi}\right].$$

$\square$

As a consequence of the above lemma, we need only consider homomorphisms of extension configurations. Indeed, the remainder of the paper is devoted to analysis of homomorphism counts of extension configurations, using Lemma 2 as the key link to relate our results back to homomorphism counts in $\mathcal{C}(\boldsymbol{d},\gamma)$.

5.2   Expected Homomorphism Counts

Now we turn to the task of estimating the expected number of homomorphisms of a fixed graph into $\mathcal{C}(\gamma, \mathbf{d})$. First we have a bit of notation to define. For two graphs labeled graphs G and H, let $Union(G, H)$ denote the set of all labeled graphs formed by selecting vertices $s_1, \ldots, s_k \in G$ and $t_1, \ldots, t_k \in H$ and identifying s_i and t_i in the disjoint union of G and H. The graphs in $Union(G, H)$ are all the graphs that can be formed by placing a copy of G on $[n]$ and then adding a (possibly overlapping) copy of H onto the same set.

We will need to carefully consider $Union(K_1, K_2)$ when K_1 and K_2 are both extension configurations of some graph H. Since $V(H)$ is a subset of the vertices of any of extension configurations, let $Union_H(K_1, K_2)$ denote the set of all graphs in $Union(K_1, K_2)$ where the copy of $v \in V(H)$ in $V(K_1)$ is matched to the copy of v in $V(K_2)$. Note that other pairs of vertices may also be identified. Finally, if we consider $Union_H(K_1, K_1)$, we wish to avoid considering the copy of K_1 in this set, i.e. the situation when every vertex in one copy of K_1 is identified with its counterpart in the other copy. Let $Union_H(K_1, K_2)^\circ$ denote $Union_H(K_1, K_2)$ if $K_1 \neq K_2$ and $Union_H(K_1, K_1) \setminus \{K_1\}$ when $K_1 = K_2$.

To begin we have the following technical lemma which is a consequence of Lemma 2 and the Bonferroni inequalities [6]. The proof can be found in the full version of the paper.

Lemma 3. *Fix a graph H. If $\mathbf{d}$ is a vector of length n and $\gamma \in (0, 1)$, then the subgraph counts of H in $\mathcal{C}(\gamma, \mathbf{d})$ are upper and lower bounded by the subgraph counts of the extension configurations of H in $\mathcal{G}(\mathbf{d})$ where the lower bound has an error term which depends on pairs of extension configurations. Formally, there are sets of constants $C(K, \gamma)$ and $C(K_1, K_2, \gamma)$ such that*

$$S_1(H) - S_2(H) \leq \mathbb{E}_{G \hookleftarrow \mathcal{C}(\gamma, \mathbf{d})} [\hom(H, G)] \leq S_1(H)$$

where

$$S_1(H) = \sum_{K \in Ext(H)} C(K, \gamma) \, \mathbb{E} [\hom(K, G_0)]$$

and

$$S_2(H) \sum_{K_1, K_2 \in Ext(H)} \sum_{L \in Union_H(K_1, K_2)^\circ} C(L, \gamma) \, \mathbb{E} [\hom(L, G_0)].$$

Now we can prove our first main result:

Theorem 3. *Fix a graph H and let Δ be the maximum degree over all graphs $L \in Union_H(K_1, K_2)$ for all pairs of extension configurations $K_1, K_2 \in Ext(H)$. If $\mathbf{d}$ is a vector obeying the (a, b, Δ)-volume growth condition for some constants $a \in [0, \frac{1}{2})$ and $b \in (0, 1 - a]$ such that $4a + b - 2 < 0$, then the expected number of graph homomorphisms of H into $G \hookleftarrow \mathcal{C}(\gamma, \mathbf{d})$ is determined up to constants by the homomorphism counts of extension configurations of H in the Chung-Lu model $\mathcal{G}(\mathbf{d})$. More formally, we have*

$$\mathbb{E}_{G \hookleftarrow \mathcal{C}(\gamma, \mathbf{d})} [\hom(H, G)] = (1 - o(1)) \sum_{K \in Ext(H)} C(K, \gamma) \, \mathbb{E} [\hom(K, G_0)].$$

Proof. Our goal is to show that

$$\frac{\mathbb{E}\left[\hom(L, G_0)\right]}{\mathbb{E}\left[\hom(K_1, G_0)\right]} = o(1)$$

for any pair of extension configurations $K_1, K_2 \in Ext(H)$ and any $L \in Union_H(K_1, K_2)^\circ$. The theorem will then follow by Lemma 3.

Fix $K_1 \in Ext(H)$, $K_2 \in Ext(H)$. Fix a choice of $L \in Union_H(K_1, K_2)^\circ$. By Lemma 1,

$$\frac{\mathbb{E}\left[\hom(L, G_0)\right]}{\mathbb{E}\left[\hom(K_1, G_0)\right]} = (1 + o(1)) \frac{\prod_{v \in L} \text{Vol}_{\deg_L(v)}(\mathbf{d})}{\prod_{v \in K_1} \text{Vol}_{\deg_{K_1}(v)}(\mathbf{d})} \frac{1}{\text{Vol}(\mathbf{d})^{e(L) - e(K_1)}}$$

Since $\Delta(L) \leq \Delta$, we may apply the (a, b, Δ)-volume growth condition as follows:

$$\frac{\mathbb{E}\left[\hom(L, G_0)\right]}{\mathbb{E}\left[\hom(K_1, G_0)\right]} = \Theta\left(\text{Vol}(\mathbf{d})^{\sum_{v \in L}(a \deg_L(v) + b) - \sum_{v \in K_1}(a \deg_{K_1}(v) + b) + e(K_1) - e(L)}\right)$$

$$= \Theta\left(\text{Vol}(\mathbf{d})^{2ae(L) + bv(L) - 2ae(K_1) - bv(K_1) + e(K_1) - e(L)}\right)$$

$$= \Theta\left(\text{Vol}(\mathbf{d})^{(2a-1)(e(L) - e(K_1)) + b(v(L) - v(K_1))}\right)$$

Consider the case when $v(L) = v(K_1)$. Since $L \in Union_H(K_1, K_2)^\circ$, there must be at least once edge in L which is not in K_1. Thus, $(2a - 1)(e(L) - e(K_1)) < 0$ as $a < \frac{1}{2}$ and we are done. Similarly if $v(L) > v(K_1)$, observe that each vertex in $L \setminus K_1$ is an external vertex in K_2, and therefore contributes at least 2 edges to L which are not in K_1. Hence if $v(L) > v(K_1)$,

$$\frac{\mathbb{E}\left[\hom(L, G_0)\right]}{\mathbb{E}\left[\hom(K_1, G_0)\right]} = \Theta\left[\text{Vol}(\mathbf{d})^{(4a+b-2)(v(L) - v(K_1))}\right].$$

Since $4a + b - 2 < 0$, we conclude that

$$\frac{\mathbb{E}\left[\hom(L, G_0)\right]}{\mathbb{E}\left[\hom(K_1, G_0)\right]} = o(1) \tag{1}$$

if $v(L) > v(K_1)$.

Recall $S_2(H)$ and $S_1(H)$ from Lemma 3. By (1), we conclude that $S_2(H) = o(S_1(H))$, and thus

$$\mathbb{E}_{G \leftarrow \mathcal{C}(\gamma, \mathbf{d})}\left[\hom(H, G)\right] = (1 + o(1))\left(\sum_{K \in Ext(H)} C(K, \gamma)\, \mathbb{E}\left[\hom(K, G_0)\right]\right)$$

by Lemma 3.   □

5.3   Concentration of Subgraph Counts

Now that we have an estimate of the expected number of homomorphisms of a fixed graph, we can also prove a concentration result via Chebyshev's inequality.

Much like Lemma 3, we will need to express the expected square of a homomorphism counts as a sum over extension configurations. Once again, the definition of a union graph will be the key step. The proof can be found in the full version of the paper.

Lemma 4. *Fix a graph H. If $\mathbf{d}$ is a vector of length n and $\gamma \in (0,1)$, then the square of the subgraph count of H in $\mathcal{C}(\gamma, \mathbf{d})$ is upper bounded by subgraph counts of extension configurations of graphs in $Union(H, H)$. More formally, there are constants $C(K, \gamma)$ such that*

$$\mathbb{E}\left[\hom(H, G)^2\right] \leq \sum_{L \in Union(H, H)} \sum_{K \in Ext(L)} C(K, \gamma)\, \mathbb{E}\left[\hom(K, G_0)\right]$$

Lemma 4 leads us to consider extension configurations of graphs in $Union(H, H)$. The following lemma shows that extension configurations of two overlapping copies of H are precisely the union of two extension configurations of H. The proof can be found in the full version of the paper.

Lemma 5. *Fix $L \in Union(H, H)$. If $K \in Ext(L)$ then K is the union of two extension configurations of H.*

Finally, the homomorphism counts of the union of two extension configurations is bounded by the product of their homomorphism counts. The proof can be found in the full version of the paper.

Lemma 6. *Let $K_1, K_2 \in Ext(H)$ be extension configurations, and fix $K \in Union(K_1, K_2)$. If $\mathbf{d}$ satisfies the $(a, b, \Delta(K))$ volume growth condition, then the ratio of subgraph counts of K in $\mathcal{G}(\mathbf{d})$ with subgraph counts of K_1 and K_2 in $\mathcal{G}(\mathbf{d})$ is controlled by the density of $K_1 \cap K_2$. More formally,*

$$\frac{\mathbb{E}\left[\hom(K, G_0)\right]}{\mathbb{E}\left[\hom(K_1, G_0)\right]\mathbb{E}\left[\hom(K_1, G_0)\right]} = (1 + o(1))\, \text{Vol}(\mathbf{d})^{(1-2a)e(K_1 \cap K_2) - bv(K_1 \cap K_2)}$$

With these lemmas in hand, we can prove our second main result.

Theorem 4. *Fix a graph H and constants $a \in [0, \frac{1}{2})$ and $b \in (0, 1 - a]$. Let $\mathbf{d}$ denote a vector obeying the $(a, b, \Delta(H))$ volume growth condition. If $\frac{e(K')}{v(K')} < \frac{b}{1-2a}$ for every subgraph $K' \subseteq K$ of every $K \in Ext(H)$, then the subgraph count of H in a random clustering graph $\mathcal{C}(\gamma, \mathbf{d})$ is concentrated around its expected value, i.e.*

$$\hom(H, G) = (1 + o(1)) \sum_{K \in Ext(H)} C(K, \gamma)\, \mathbb{E}_{G_0 \leftarrow \mathcal{G}(\mathbf{d})}\left[\hom(K, G_0)\right]$$

with high probability over the choice of $G \leftarrow \mathcal{C}(\gamma, \mathbf{d})$.

Proof. We aim to use Chebyshev's inequality. Let $\sigma(n)$ denote a function such that $\lim_{n \to \infty} \sigma(n) = \infty$ to be determined later, and let $\overline{E} = \mathbb{E}\left[\hom(H, G)\right]$ for brevity. By Chebyshev's inequality,

$$\mathbb{P}\left[|\hom(H, G) - \overline{E}| > \frac{\overline{E}}{\sigma(n)}\right] < \sigma(n)^2 \left(\frac{\mathbb{E}\left[\hom(H, G)^2\right]}{\overline{E}^2} - 1\right) \tag{2}$$

By Lemmas 4 and 5, we may rewrite the sum as follows:

$$\mathbb{E}\left[\hom(H,G)^2\right] \leq \sum_{L \in Union(H,H)} \sum_{K \in Ext(L)} C(K,\gamma)\,\mathbb{E}\left[\hom(K,G_0)\right]$$

$$\leq \sum_{K_1,K_2 \in Ext(H)} \sum_{K \in Union(K_1,K_2)} C(K,\gamma)\,\mathbb{E}\left[\hom(K,G_0)\right]$$

By Lemma 6, for each $K_1, K_2 \in Ext(H)$ and $K \in Union(K_1, K_2)$ we have

$$\mathbb{E}\left[\hom(K,G_0)\right] \leq \mathrm{Vol}(\mathbf{d})^{\beta(a,b,K_1 \cap K_2)}\,\mathbb{E}\left[\hom(K_1,G_0)\right]\mathbb{E}\left[\hom(K_2,G_0)\right]$$

where $\beta(a,b,K_1 \cap K_2) = (1-2a)e(K_1 \cap K_2) - bv(K_1 \cap K_2)$. Observe that $K_1 \cap K_2$ is a subgraph of K_1 and K_2, and the maximum density of any nontrivial subgraph of K_1 or K_2 is strictly less that $\frac{d}{1-2c}$. Thus, it follows that $\beta(a,b,K_1 \cap K_2) \leq 0$, and equality holds if and only if $K_1 \cap K_2$ is the empty graph, i.e. $K = K_1 \sqcup K_2$. Hence we may combine all terms except those corresponding to $K_1 \sqcup K_2$ into the $o(1)$ term as follows:

$$\mathbb{E}\left[\hom(H,G)^2\right] \leq \sum_{K_1,K_2 \in Ext(H)} (1+o(1))C(K_1 \sqcup K_2,\gamma)\,\mathbb{E}\left[\hom(K_1,G_0)\right]\mathbb{E}\left[\hom(K_2,G_0)\right]$$

Recall from Lemma 3 that $C(K_1 \sqcup K_2,\gamma)$ is a conditional probability which depends on coin flips for paths in K_1 and K_2. We note that $C(K_1 \sqcup K_2,\gamma) = C(K_1,\gamma)C(K_2,\gamma)$ as there are no edges shared between K_1 and K_2 in their disjoint union. Therefore,

$$\mathbb{E}\left[\hom(H,G)^2\right] \leq (1+o(1))\,\mathbb{E}\left[\hom(H,G)\right]^2 = (1+o(1)\overline{E}^2 \qquad (3)$$

by Theorem 3.

Combining (2) and (3), we find that

$$\mathbb{P}\left[\left|\hom(H,G) - \overline{E}\right| > \frac{\overline{E}}{\sigma(n)}\right] < \sigma(n)^2 \left(\frac{(1+o(1))\overline{E}^2}{\overline{E}^2} - 1\right)$$

$$= o(\sigma(n)^2)$$

By choosing σ growing sufficiently slow so that the above expression is $o(1)$, the theorem follows. □

Finally, we can apply Theorem 4 to compute the clustering coefficient of $\mathcal{C}(\gamma,\mathbf{d})$ and also show concentration.

Theorem 5. *If $\mathbf{d}$ is a vector obeying a power-law with exponent $2.5 \leq \beta < 3$ then with high probability the clustering coefficient of a random clustering graph $G \hookleftarrow \mathcal{C}(\gamma,\mathbf{d})$ is $(1+o(1))\gamma$.*

Proof. We begin by verifying the hypothesis of Theorems 3 and 4. From Example 2, $\mathbf{d}$ satisfies the $(1/\beta, 1/\beta, \infty)$ volume growth condition. To apply Theorem 3, we must check that $\frac{4}{\beta} + \frac{1}{\beta} < 2$, which holds for $\beta > 2.5$. To apply Theorem 4, we must check that $e(H)/v(H) < \frac{\frac{1}{\beta}}{1-\frac{2}{\beta}}$ for each subgraph $H \subseteq K$ where K is an extension configuration of P_2 or K_3. From Fig. 4, the extension configurations of P_2 are P_2, P_3, P_4, and $K_{1,3}$. From Fig. 3, the extension configurations of K_3

are K_3, P_2, $K_{1,3}$, C_4, C_5, C_6. One can easily verify that $e(H)/v(H) \leq 1$ for any subgraph H of these graphs. Since $\frac{\frac{1}{\beta}}{1-\frac{2}{\beta}} = \frac{1}{\beta-2} > 1$ for $2.5 < \beta < 3$, we may apply Theorem 4 and conclude that

$$\hom(K_3, G) = (1 + o(1)) \sum_{K \in Ext(K_3)} C(K, \gamma) \, \mathbb{E}\left[\hom(K, G_0)\right] \tag{4}$$

$$\hom(P_2, G) = (1 + o(1)) \sum_{K \in Ext(P_2)} C(K, \gamma) \, \mathbb{E}\left[\hom(K, G_0)\right] \tag{5}$$

each with probability $1 - o(1)$. We now estimate both sums. By Lemma 1 and the $(1/\beta, /\beta, \infty)$ volume growth condition, we have the following expected homomorphism counts.

K	P_2	$K_{1,3}$	C_5	C_6
$\mathbb{E}\left[\hom(K, G_0)\right]$	$\mathrm{Vol}(\mathbf{d})^{\frac{3}{\beta}-1}$	$\mathrm{Vol}(\mathbf{d})^{\frac{4}{\beta}}$	$\mathrm{Vol}(\mathbf{d})^{5(\frac{3}{\beta}-1)}$	$\mathrm{Vol}(\mathbf{d})^{6(\frac{3}{\beta}-1)}$

The two largest terms are $\mathrm{Vol}(\mathbf{d})^{\frac{4}{\beta}}$ and $\mathrm{Vol}(\mathbf{d})^{6(\frac{3}{\beta}-1)}$, and since $\frac{4}{\beta} > 6(\frac{3}{\beta}-1)$ for $2.5 < \beta < 3$, it follows that $K_{1,3}$ dominates the sum (4). We apply the same analysis to P_2 and find that

K	P_2	P_3	P_4	$K_{1,3}$
$\mathbb{E}\left[\hom(K, G_0)\right]$	$\mathrm{Vol}(\mathbf{d})^{\frac{3}{\beta}}$	$\mathrm{Vol}(\mathbf{d})^{\frac{6}{\beta}-1}$	$\mathrm{Vol}(\mathbf{d})^{\frac{9}{\beta}-2}$	$\mathrm{Vol}(\mathbf{d})^{\frac{4}{\beta}}$

Again $\frac{4}{\beta}$ is the largest exponent, so $K_{1,3}$ also dominates the sum (5). From Lemma 3, we find that the constants $C(K_{1,3}, \gamma)$ in Equations (4) and (5) are γ^3 and γ^2 respectively. Hence,

$$C_1(G) = \frac{\hom(K_3, G)}{\hom(P_2, G)} = \frac{(1 + o(1))\gamma^3 \, \mathrm{Vol}_3(\mathbf{d})}{(1 + o(1))\gamma^2 \, \mathrm{Vol}_3(\mathbf{d})} = (1 + o(1))\gamma$$

with probability $1 - o(1)$. $\qquad\qquad\qquad\qquad\qquad\qquad\qquad\qquad\qquad\qquad\qquad\square$

In conclusion, there are many future directions which we do not consider here. The omitted proofs can be found in the full version of the paper.

Acknowledgments. We thank the anonymous referees for their thorough reviews and invaluable suggestions.

References

1. Aiello, W., Bonato, A., Cooper, C., Janssen, J., Prałat, P.: A spatial web graph model with local influence regions. Internet Math. **5**(1–2), 175–196 (2008)
2. Aiello, W., Chung, F., Lu, L.: A random graph model for power law graphs. Exp. Math. **10**(1), 53–66 (2001). https://doi.org/em/999188420
3. Aiello, W., Chung, F., Lu, L.: Random evolution in massive graphs. In: Abello, J., Pardalos, P.M., Resende, M.G.C. (eds.) Handbook of Massive Data Sets. MC, vol. 4, pp. 97–122. Springer, Boston, MA (2002). https://doi.org/10.1007/978-1-4615-0005-6_4
4. Albert, R., Barabási, A.L.: Emergence of scaling in random networks. Science **74**(5439), 509–512 (1999)
5. Albert, R., Barabási, A.L., Jeong, H.: Scale-free characteristics of random networks: the topology of the world-wide web. Phys. A: Stat. Mech. Appl. **281**, 69–77 (2000). https://doi.org/10.1016/S0378-4371(00)00018-2
6. Alon, N., Spencer, J.H.: The Probabilistic Method. John Wiley & Sons, Hoboken (2016)
7. Bradonjić, M., Hagberg, A., Percus, A.G.: The structure of geographical threshold graphs. Internet Math. **5**(1–2), 113–139 (2008)
8. Bringmann, K., Keusch, R., Lengler, J.: Geometric inhomogeneous random graphs. Theor. Comput. Sci. **760**, 35–54 (2019)
9. Chung, F., Lu, L.: The average distance in a random graph with given expected degrees. Internet Math. **1**(1), 91–113 (2004). https://doi.org/10.1080/15427951.2004.10129081
10. Chung, F., Lu, L.: The volume of the giant component of a random graph with given expected degrees. SIAM J. Discrete Math. **20**(2), 395–411 (2006)
11. Chung, F., Lu, L., Vu, V.: The spectra of random graphs with given expected degrees. Internet Math. **1**(3), 257–275 (2004)
12. Erdős, P., Rényi, A.: On the evolution of random graphs. Magyar Tud. Akad. Mat. Kutató Int. Közl. **5**, 17 (1960)
13. Hyland-Wood, D., Carrington, D., Kaplan, S.: Scale-free nature of java software package, class and method collaboration graphs. In: Proceedings of the 5th International Symposium on Empirical Software Engineering. Citeseer (2006)
14. Jacob, E., Mörters, P.: A spatial preferential attachment model with local clustering. In: Bonato, A., Mitzenmacher, M., Prałat, P. (eds.) WAW 2013. LNCS, vol. 8305, pp. 14–25. Springer, Cham (2013). https://doi.org/10.1007/978-3-319-03536-9_2
15. Krioukov, D., Papadopoulos, F., Kitsak, M., Vahdat, A., Boguná, M.: Hyperbolic geometry of complex networks. Phys. Rev. E **82**(3), 036106 (2010)
16. Newman, M.E.: The structure and function of complex networks. SIAM Rev. **45**(2), 167–256 (2003)
17. Steyvers, M., Tenenbaum, J.B.: The large-scale structure of semantic networks: Statistical analyses and a model of semantic growth. Cogn. Sci. **29**(1), 41–78 (2005)

Topological Analysis of Temporal Hypergraphs

Audun Myers[1]([✉]), Cliff Joslyn[1], Bill Kay[2], Emilie Purvine[1], Gregory Roek[1],
and Madelyn Shapiro[1]

[1] Pacific Northwest National Laboratory, Mathematics of Data Science,
Richland, USA
audun.myers@pnnl.gov
[2] Pacific Northwest National Laboratory, Computational Mathematics,
Richland, USA

Abstract. In this work we study the topological properties of temporal hypergraphs. Hypergraphs provide a higher dimensional generalization of a graph that is capable of capturing multi-way connections. As such, they have become an integral part of network science. A common use of hypergraphs is to model events as hyperedges in which the event can involve many elements as nodes. This provides a more complete picture of the event, which is not limited by the standard dyadic connections of a graph. However, a common attribution to events is temporal information as an interval for when the event occurred. Consequently, a temporal hypergraph is born, which accurately captures both the temporal information of events and their multi-way connections. Common tools for studying these temporal hypergraphs typically capture changes in the underlying dynamics with summary statistics of snapshots sampled in a sliding window procedure. However, these tools do not characterize the evolution of hypergraph structure over time, nor do they provide insight on persistent components which are influential to the underlying system. To alleviate this need, we leverage zigzag persistence from the field of Topological Data Analysis (TDA) to study the change in topological structure of time-evolving hypergraphs. We apply our pipeline to both a cyber security and social network dataset and show how the topological structure of their temporal hypergraphs change and can be used to understand the underlying dynamics.

1 Introduction

Complex networks are a natural tool for studying dynamical systems where elements of the system are modeled in a dyadic way and evolve over time. There are many real-world examples, such as social networks [28], disease spread dynamics [18], manufacturer-supplier networks [31], power grid networks [26], and transportation networks [9]. The underlying complex dynamical systems driving these

Information release number: PNNL-SA-181478.

networks cause temporal changes to their structure, with connections and elements added and removed as the dynamical system changes. We can summarize this category of complex network as dynamical networks [17] where the resulting graph is a temporal graph with temporal attributes associated to each connection and/or element of the complex network.

While temporal networks are useful in understanding systems with dyadic relations between elements, the complex network is not always satisfactory for modeling the relationship between multiple entities [11]. For data with multi-way relations that cannot be described by dyadic connections, hypergraphs capture richer information about community structure. For example, in Sect. 3.1 we explore a hypergraph built from Reddit data (PAPERCRANE [5]) on threads about COVID-19. A dyadic model, where an edge links two users if and only if they posted in the same thread, loses all information about thread size. In contrast, a hypergraph, where each thread is an edge and a user is in a thread if and only if they posted in that thread, retains the total structure of the data. In this way, hypergraph analytics are a powerful tool when higher order structure is of interest. Some instances where hypergraphs have been useful include human gene sets [12,19] where genes interact in complex combinations, cyber data [19] with the domain name systems mapping multiple domains and IPs, and social networks with interactions between large groups [11].

In many use cases, individual snapshots of a complex system are less important than analysis of how the system *changes*. Often, these networks are further improved by modeling them as Temporal HyperGraphs (THG) in the same way as temporal graphs, with temporal attributes (e.g., intervals or times) associated to the multi-way connections and elements. Examples can be found in many settings, such as anomaly detection in automotive data (CAN bus) [16] and cybersecurity using the Operationally Transparent Cybersecurity data we consider in Sect. 3.2 [15].

Many common tools for studying the characteristics of THGs are based on summary statistics of the underlying hypergraph snapshots. These statistics provide insight to dynamic changes in the structure of the underlying hypergraph. For example, in [8], Cencetti *et al.* studied temporal social networks as hypergraphs and were able to measure the burstiness of multi-way social interactions using a burstiness statistic. While statistics such as this can be informative for change detection and insights into the system dynamics, they are lacking in their ability to interpret changes in the structure of the temporal hypergraph. Another approach for studying temporal hypergraphs is through visual analytics. In [13], a matrix based visual analytics tool was designed for temporal hypergraph analysis which provides insights into the dynamic changes of the hypergraph. However, visualization tools are naturally limited in their ability to be automatically interpreted and often require expertise to properly understand.

What distinguishes hypergraphs from graphs is that hyperedges come not only in arbitrary sizes, but also connected into arbitrarily complex patterns.

As such, they can actually have a complex mathematical topology[1] as complex "gluings" of multi-dimensional objects which can have a complex shape and structure. Studying the topology of hypergraphs is a becoming an increasingly large area, frequently exploiting their representation as Abstract Simplicial Complexes (ASCs).

The field of Topological Data Analysis (TDA) [10,33] aims to measure the shape of data. Namely, data is represented as an ASC, whose homology is then measured to detect overall topological shape, including components, holes, voids, and higher order structures. However, this often requires the choice of parameters to generate the ASC from the data, which is typically nontrivial. Another, more automatic, approach for measuring the shape of data is to use persistent homology from TDA. This method for studying the shape of data extracts a sequence of ASCs from the data, which is known as a filtration. Persistent homology has been successfully applied to a wide application domains, including manufacturing [20,32], biology [4], dynamical systems [22,29], and medicine [27]. Many of the applications either represent the data as point clouds or graphs. For point cloud data, filtrations are commonly generated as a collection of Vietoris-Rips complexes [10] determined by identifying points within a Euclidean distances of an increasing radius parameter. For graph data a similar process would be to use a distance filtration with the shortest path distance [3,22].

Hypergraphs have also been studied using tools from TDA. Namely, the work in [14] shows how the homology of hypergraphs can be studied using various ASC representations such as the associated ASC [25] or the relative/restricted barycentric subdivision.

However, a requirement for applying persistent homology is that there is a monotonic inclusion mapping between subsequent members of a sequence of ASCs (i.e., each subsequent ASC in the sequence has its previous as a subset). Many sequences of ASCs associated with data sets are not monotonic, however, we still want to track their changing structure. This is commonly true for temporal data, where, for example, hypergraph edges can appear and then disappear over time, which would break the monotonicity requirement for persistent homology.

To solve this problem, zigzag persistence [7] can be applied. Instead of measuring the shape of static point cloud data through a distance filtration (e.g., a Vietoris-Rips filtration), zigzag persistence measures how long a topology generator persists in a sequence of ASCs by using a an alternating sequence of ASCs, called a "zigzag filtration".

Both PH and zigzag persistence track the formation and disappearance of the homology through a persistence diagram or barcode as a two-dimensional summary consisting of persistence pairs (b, d), where b is the birth or formation time of a generator of a "hole" of a certain dimension, and d is its death or disappearance time. For example, in [30] the Hopf bifurcation is detected through zigzag persistence of Vietoris-Rips complexes over sliding windows using

---

[1] Notice we use "topology" here in the formal sense, as distinct from how this is used informally in graph applications to refer to connectivity patterns in networks.

the one-dimensional homology. Another recent application [23] studies temporal networks, where graph snapshots were represented as ASCs using the Vietoris-Rips complex with the shortest path distance. However, both of these methods require a distance parameter to be selected to form the ASC at each step, which is typically not a trivial choice.

The resulting persistence barcodes from zigzag persistence can also be vectorized using methods such as persistence images [1] or persistence landscapes [6]. This allows for the resulting persistence diagrams to be analyzed in automatic methods using machine learning for classification or regression.

In this work we leverage zigzag persistence to study THGs. By measuring the changing structure of the temporal hypergraph through an ASC representation of the hypergraph, we are able to detect the formation, combination, and separation of components in the hypergraph as well as higher dimensional features such as loops or holes in the hypergraph and voids. The detection of these higher dimensional features is critical for temporal hypergraph analysis as they may be of consequence depending on the application domain. Additionally, in comparison to creating an abstract ASC from point cloud or graph data, no distance parameter needs to be chosen as there are natural representations of a hypergraph as an ASC [14].

In Sect. 2 of this paper we introduce THGs and an ASC representation of hypergraphs. We then overview persistent homology and zigzag persistence. In Sect. 3 we demonstrate how our method can be applied to two data sets drawn from social networks and cyber data. Lastly, in Sect. 4 we provide conclusions and future work.

2   Method and Background

In this section the method for studying temporal hypergraphs using zigzag persistence is developed alongside the necessary background material. Our method is a confluence of zigzag persistence and the ASC representation of hypergraphs for the topological analysis of THGs. Namely, we develop a pipeline for applying zigzag persistence to study changes in the shape of a temporal hypergraph using a sliding window procedure. This pipeline is outlined in Fig. 1.

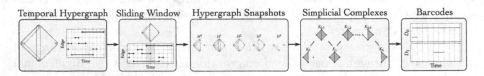

Fig. 1. Pipeline for applying zigzag persistence to temporal hypergraphs.

We begin with a temporally edge-attributed hypergraph in Fig. 1–*Temporal Hypergraph*, where each edge has active intervals associated to it as described in

Sect. 2.1. Next, we use a Fig. 1–*Sliding Window* procedure, where we choose a window size w and shift s that is slid along the time domain of the set of intervals in discrete steps. Using each sliding window, we generate Fig. 1–*Hypergraph Snapshots* at each window, which is described in Sect. 2.2. We then represent each snapshot as a Fig. 1–*ASC* using the associated ASC in Sect. 2.3. Next, we introduce simplicial homology for studying the shape of an ASC in Sect. 2.4. This leads to the method for relating the homology within a sequence of ASCs known as zigzag persistent homology in Sect. 2.5, which is used for calculating the persistent homology of the temporal hypergraph represented as a barcode of persistent diagram (Fig. 1–*Barcodes*).

To illustrate our procedure we provide a simple example throughout each step in the pipeline. For the example and the remaining results we use the Python packages HyperNetX[2] to generate the hypergraphs and Dionysus2[3] to calculate the zigzag persistence.

2.1 Temporal Hypergraphs

A graph $G(V, E)$ is composed of a set of vertices connected using a set of edges with $E \subseteq \binom{V}{2}$. A hypergraph $H(V, E)$ is composed of a set of vertices V and a family of edges E, where for each $E_i \in E, E_i \subseteq V$. In this way a hypergraph can capture a connection between k vertices as a k-edge. For example, consider the toy hypergraph in Fig. 2a with four nodes $V = \{A, B, C, D\}$ and five hyperedges $E = \{E_1, E_2, E_3, E_4, E_5\}$. These hyperedges in the example range in size from edge $E_2 = (D)$ as a 1-edge to edge $E_4 = (A, B, C)$ as a 3-edge.

A temporal hypergraph $H(V, E, T)$ is a replica of its underlying static hypergraph with the addition of temporal attributes T associated to either the vertices, edges, or incidences. An attribute to an incidence occurs when the temporal information associated to a node is relative to the hyperedge. In this work we only use temporal information attributed to the edges. However, our pipeline could

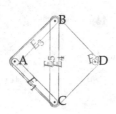
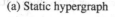

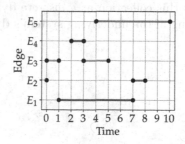

(a) Static hypergraph

(b) Temporal information stored as points and intervals for each edge.

Fig. 2. Toy example temporal hypergraph.

[2] HyperNetX: https://pnnl.github.io/HyperNetX.
[3] Dionysus2: https://mrzv.org/software/dionysus2/.

be adapted to any or all of the three temporal attribution types. Returning to our toy example hypergraph H in Fig. 2a, we include temporal information as a set of intervals associated to the time when each edge is active (e.g., E_2 is active for the point interval $[0, 0]$ and interval $[7, 8]$).

2.2 Sliding Windows for Hypergraph Snapshots

The sliding window procedure is a ubiquitous part of signal processing, in which a time series or signal is segmented into discrete windows that slide along its time domain. Specifically, Given a time domain $[t_0, t_f]$, window size w, and shift s, we create a set of windows that cover the time domain interval as

$$\mathcal{W} = \{[t_0, t_0 + w], [t_0 + s, t_0 + s + w], [t_0 + 2s, t_0 + 2s + w], \ldots, [t_0 + \ell s, t_0 + \ell s + w]\},$$
(1)

The window size and shift should be such that $s \leq w$. In this way the union of all windows covers the entire domain and adjacent windows do not have a null intersection.

For each sliding window $W_i \in \mathcal{W}$ we create a sub-hypergraph snapshot using an intersection condition between the sliding window interval W_i and the collection of intervals associated to each edge in the temporal hypergraph. The intervals are considered closed intervals in this work. This procedure is done by including an edge if there is a nonempty intersection between the edge's interval set and the sliding window interval W_i. We formalize this as

$$H_i = \{E_j \in E \mid I(E_j) \cap W_i \neq \emptyset\},$$
(2)

where $E_j \in E$ is an edge in the set of edges of the static hypergraph and $I(E_j)$ is the interval collection for edge E_j. The resulting sub-hypergraph snapshot collection of $\mathcal{W}$ is

$$\mathcal{H} = \{H_0, H_1, \ldots, H_t, \ldots, H_\ell\}.$$

We can cast this collection as a discrete dynamical process $H_t \mapsto H_{t+1}$ to gain understanding of the underlying system's dynamics.

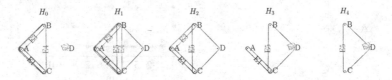

Fig. 3. Sequence of sub-hypergraphs $\mathcal{H}$ from the sliding window procedure with corresponding ASCs.

To demonstrate the sliding window procedure for getting hypergraph snapshots we use the toy example temporal hypergraph from Fig. 2 and window parameters $w = 2$ and $s = 2$. Using these parameters we get the sliding windows as

$$\mathcal{W} = \{[0,2], [2,4], [4,6], [6,8], [8,10]\}.$$

Hypergraphs from each window are generated as subsets of the static H depending on the overlap of the window and the activity intervals associated to each edge. For example, window $W_2 = [4,6]$ has the hypergraph H_2 with edges $\{E_1, E_3, E_5\}$ based on the overlap between W_2 and the collection of intervals of each edge shown in Fig. 2b. Additionally, each hypergraph has now both an index and time associated to it. The index is as was previously stated (e.g., H_2 has index 2) and the time is the average time of the corresponding window (e.g., W_2 has an average time of $(4+6)/2 = 5$). Applying this hypergraph snapshot procedure using the sliding windows we get the five hypergraphs shown in Fig. 3.

2.3 Associated ASC of a Hypergraph

An ASC K is a collection of simplices, with a simplex $\sigma \subseteq P$ as a subset of n points from a set of points P and simplex dimension $n - 1$. This results in points (1-edge) as 0-simplices, lines (2-edge) as 1-simplices, triangles (3-edge) as 2-simplices, etc. We denote the simplex σ as a face if $\sigma \subseteq \tau$ with τ as another simplex. Additionally, a simplex σ of dimension $n - 1$ is required to be closed under face relation, which is all of its subsimplices (faces) as the power set of the simplex. The dimension of an ASC is the dimension of the largest simplex. ASCs are often used to represent geometric structures and as such are referred to as geometric simplicial complexes. However, we can also refer to them as abstract simplicial complexes for purely combinatorially purposes.

We can generate the associated ASC of a hypergraph [25] using the simplices associated to each hyperedge and building the closure under face relations, which is the power set of each hyperedge.

To apply zigzag persistence to study the changing topology of our hypergraph snapshots, we need to first represent our collection of hypergraph snapshots $\mathcal{H}$ as a sequence of ASCs $\mathcal{K}$ which will later be used to create the zigzag persistence module. While there are several methods for representing a hypergraph as an ASC [14], we leverage an adaptation of the associated ASC method from [25]. The associated ASC of a hypergraph H is defined as

$$K(H) = \{\sigma \in \mathcal{P}(E_i) \setminus \emptyset \mid E_i \in E\}, \tag{3}$$

where E is the edge set of the hypergraph H, $E_i \in E$, and $\mathcal{P}(E_i)$ is the power set of E_i. Equation 3 provides a first starting point for calculating the zigzag persistence, however, it is computationally cumbersome. Specifically, for a large k-edge the computational requires

$$\sum_{j=0}^{k} \binom{k+1}{j+1} = 2^{k+1} - 1$$

subsimplices. However, the computation of homology of dimension p only requires simplices of size $p+1$ to be included in the ASC. As such, we define the *modified associated ASC* as

$$K(H, p) = \{\sigma \in \mathcal{P}_{p+1}(E_i) \setminus \emptyset \mid E_i \in E\}, \tag{4}$$

where $\mathcal{P}_{p+1}$ is the modified power set to only include elements of the set up to size $p + 1$ or $\binom{E_i}{p+1}$. The modified associated ASC reduces the computational demand by only requiring

$$\sum_{j=0}^{p+1} \binom{k+1}{j+1}$$

subsimplices for a k-edge.

Applying Eq. (4) to each hypergraph in $\mathcal{H}$ allows us to get a corresponding sequence of ASCs as $\mathcal{K}$. For the hypergraph snapshots $\mathcal{H}$ shown in Fig. 3 the modified associated ASCs $\mathcal{K}$ are shown in Fig. 4.

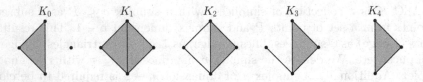

Fig. 4. Sequence of associated ASCs from hypergraph snapshots in Fig. 3.

2.4   Simplicial Homology

Simplicial homology is an algebraic approach for studying the shape of an ASC by counting the number of p-dimensional holes, where $p = 0$ are connected components, $p = 1$ are graph triangles, $p = 2$ are three-dimensional hollow tetrahedrons, and so on. We can represent the collection of p-dimensional holes of an ASC K as the Betti vector $\beta(K) = [b_0, b_1, b_2, \ldots]$, where b_p is the number of p-dimensional holes known as a Betti number. In this work we do not overview the details on how the Betti numbers are calculated, but we direct the reader to [21,24] for a formal introduction.

By calculating the Betti numbers for our sequence of ASCs in Fig. 4, we get the Betti vectors in Fig. 5. These Betti numbers are informative on the changing topology of the hypergraph snapshots in Fig. 3; however, they do not capture information on how the topology between the snapshots are related.

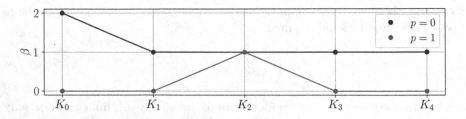

Fig. 5. Betti numbers for ASCs in Fig. 4.

For example, by observation of the hypergraph snapshots we know that there is one main component that persists through the entire sequence of ASCs, but this information can not be known directly from the Betti numbers. The Betti numbers do not tell the complete story of this component persisting the whole time. While they do tell us there is at least one component in each snapshot, these components do not necessarily need to be the same component in each snapshot to get the same Betti vectors. As such, we need to use a method to track how the homology is changing and related between the sequence of ASCs. To do this we implement zigzag persistent homology.

2.5   Zigzag Persistent Homology

This section provides a conceptual introduction to persistent homology and how it generalizes to zigzag persistent homology. We suggest [21,24] for a detailed introduction on persistent homology.

Persistent homology [33], a filtration tool from the field of Topological Data Analysis (TDA) [10,33], is used to gain a sense of the shape and size of a dataset at multiple dimensions and filtration values. For example, it can measure connected components (dimension zero), holes (dimension one), voids (dimension two), and higher dimensional analogues, as well as an idea of their general size or geometry. Persistent homology measures these shapes using a parameterized filtration to detect when homology groups are born (appear) and die (disappear).

To compute persistent homology a parameterization function is applied to the dataset to create a nested sequence of ASCs

$$K_0 \subseteq K_1 \subseteq K_2 \subseteq \ldots \subseteq K_n. \tag{5}$$

We can then calculate the homology of dimension p for each complex, $H_p(K_i)$, which is a vector space representing the p-dimensional structure of the space such as components, holes, voids, and higher dimensional features. However, this information does not yet yield how the homology of each ASC is related to the next ASC. To get this information, persistent homology uses the inclusions on the ASCs to induce linear maps on the vector spaces resulting in a construction called a persistence module $\mathcal{V}$:

$$H_p(K_{\alpha_0}) \hookrightarrow H_p(K_{\alpha_1}) \hookrightarrow H_p(K_{\alpha_2}) \hookrightarrow \ldots \hookrightarrow H_p(K_{\alpha_n}), \tag{6}$$

where $\hookrightarrow$ are the maps induced by the inclusion map between ASCs. It should be noted that in the sequence of ASCs, each vertex must be unique and consistently identified.

The appearance and disappearance of classes at various dimensions in this object can be tracked, resulting in a summary known as a persistence barcode (alternatively a persistence diagram) $\mathcal{D} = \{D_0, D_1, \ldots, D_p\}$. For each homology generator which appears at K_b and disappears at K_d, we draw an interval $[b, d]$ in the barcode. Taken together is the persistence barcode, which is the collection of persistence intervals (also called persistence pairs in the persistence diagram).

This persistent homology framework can be applied to study hypergraphs directly where a persistence module $\mathcal{V}$ is generated from a hypergraph, as described in [25], by generating a sequence of subset ASC representations of a hypergraph. However, a limitation of persistent homology is it requires each subsequent ASC to be a subset of the previous ASC to form the persistence module as shown in Eq. (5), which means at each step we are not allowed to remove simplices in the next ASC. There are many cases of real-world applications where we have a parameterized sequence of ASCs where simplices can both enter and exit the complex throughout the sequence. To alleviate this issue zigzag persistence [7] can be applied, which allows for arbitrary subset directions in the ASC sequence:

$$K_0 \leftrightarrow K_1 \leftrightarrow K_2 \leftrightarrow \ldots \leftrightarrow K_n, \tag{7}$$

where $\leftrightarrow$ denotes one of the two inclusion maps: $\hookrightarrow$ or $\hookleftarrow$. A common special case of this definition is where the left and right inclusions alternate or zigzag. For most data analysis applications using zigzag persistent we artificially construction a sequence of ASCs taking this form by interweaving the original ASCs with either unions or intersections of adjacent ASCs. For example, in Fig. 6a we use the union between the associated ASCs of the original hypergraph snapshots from Fig. 3. This sequence of interwoven ASCs fulfills the criteria of the zigzag inclusion map directions as

$$K_0 \hookrightarrow K_{0,1} \hookleftarrow K_1 \hookrightarrow K_{1,2} \hookleftarrow K_2 \hookrightarrow \ldots \hookleftarrow K_{\ell-1} \hookrightarrow K_{\ell-1,\ell} \hookleftarrow K_\ell. \tag{8}$$

for unions or

$$K_0 \hookleftarrow K_{0,1} \hookrightarrow K_1 \hookleftarrow K_{1,2} \hookrightarrow K_2 \hookleftarrow \ldots \hookrightarrow K_{\ell-1} \hookleftarrow K_{\ell-1,\ell} \hookrightarrow K_\ell \tag{9}$$

for intersections, where $K_{i,i+1} = K_i \cup K_{i+1}$.

The inclusion maps are extended to linear maps between homology groups resulting in the zigzag persistence module tracking the changing homology of Eq. (8) or (9) just as was the case for standard persistent homology. Focusing on the case of the union, the zigzag persistent homology module is

$$H_p(K_0) \hookrightarrow H_p(K_{0,1}) \hookleftarrow H_p(K_1) \hookrightarrow H_p(K_{1,2}) \hookleftarrow H_p(K_2) \hookrightarrow \ldots \hookleftarrow H_p(K_{n-1})$$
$$\hookrightarrow H_p(K_{n-1,n}) \hookleftarrow H_p(K_n). \tag{10}$$

The same algebra leveraging the linear maps between homology groups to track persistence pairs for a standard filtration in persistent homology makes it possible to compute where (when) homology features are born and die based on the zigzag persistence module, however some of the intuition is lost. Namely, we can again track the persistent homology using a persistence diagram $D = \{D_0, D_1, \ldots, D_p\}$ consisting of half-open intervals (persistence pairs) $[b, d)$; however, we now use the indices of the ASCs as the birth and death times instead of the filtration parameter. For example, if there is one-dimensional homology (i.e., a loop) that appears at K_2 and persists until it disappears at K_3, we represent this as the persistence pair (2,3). In the case of a class appearing or disappearing at the union (or intersection) complex $K_{i,i+1}$, we use the half index pair $i, i + 1$. If a topological feature persists in the last ASC in the zigzag persistence module we set its death past the last index with the pair $\ell, \ell + 1$, where ℓ is the number of ASCs (without interwoven unions or intersections).

To demonstrate how zigzag persistence tracks the changing topology in a sequence of ASCs we use a simple sequence of ASCs in Fig. 4, which were derived from the toy example in Fig. 2 using a sliding window procedure outlined in Sect. 2.2. As a first example of the application of zigzag persistence to study temporal hypergraphs we return to our toy example. We used the unions between ASCs to get the ASCs shown as $[K_0, K_{0,1}, K_1, \ldots, K_{3,4}, K_4]$ in Fig. 6a and the resulting zigzag persistence barcodes in Fig. 6b. For this example we are only investigating the topological changes in dimensions 0 and 1 since there are no higher dimensional features. There are two main changes in the homology of the ASCs that are captured in the persistence barcodes. For dimension 0, we are tracking the connected components and how they relate. At K_0 we have two connected components (the 2-simplex as the triangle and 0-simplex as the point). As such, we set the birth of the two components at the index which they both appear: 0. Next, at $K_{0,1}$ the components combine as two conjoined 2-simplices. The joining of components forces one of the components to die while the other persists; the smaller of the two components dies (the 0-simplex) dies at the index $0, 1$ with persistence interval $(0, (0, 1))$ shown in the D_0 barcode of Fig. 6b. The combined component never separates or combines with another component again and therefor it persists for the remaining persistence module finally dying after K_4 or index $4, 5$ (shown as the dashed red line) having the persistence interval $(0, (4, 5))$ in D_0. Moving to dimension 1, we are now interested in showing how the persistence barcode captures the formation and disappearance of loops in the persistence module. A loop is first formed in K_2 and persists until K_3. Therefor, this feature is represented as the persistence interval $(2, 3)$ in D_1 of Fig. 6b. This example highlights how zigzag persistence captures changes in the topology of a sequence of ASCs.

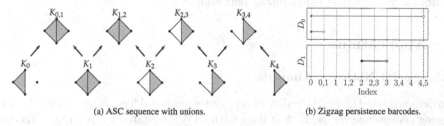

(a) ASC sequence with unions.   (b) Zigzag persistence barcodes.

Fig. 6. Zigzag persistence module and resulting barcodes for dimensions 0 and 1 for toy example introduced in Fig. 2.

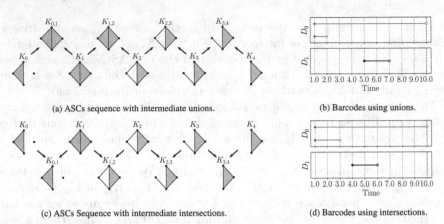

(a) ASCs sequence with intermediate unions.

(b) Barcodes using unions.

(c) ASCs Sequence with intermediate intersections.

(d) Barcodes using intersections.

Fig. 7. Sequence of ASCs from the sliding window hypergraph snapshots for both union and intersections. Zigzag persistence barcodes for temporal hypergraph example with time associated ASCs.

In this work we are interested in the analysis of temporal hypergraphs, and as such we instead want to have the barcodes track the time at which homology appears and disappears instead of the indices. To do this we substitute the index for the average time of the window associated to each ASC as shown in Fig. 7. For the intermediate ASCs (unions or intersections) we use the average time of the two windows. The only difference between the ASC sequence in Fig. 6b and Fig. 7b is that Fig. 7b has the times from the windows associated to the ASCs when computing the zigzag persistence. As such, the persistence barcode has time on the horizontal axis with the two intervals in D_0 and one in D_1 having the same sources (generators) as described in Fig. 6b.

The resulting barcodes in Fig. 7 shows that both the intersection and union methods for interweaving ASCs provide similar barcodes. We also found this same result when applying zigzag persistence to the data sets studied in this work. For the remainder of this work we will use the union method for studying temporal hypergraphs using zigzag persistence.

3   Applications

3.1   Social Network Analysis

To demonstrate the functionality of analyzing temporal hypergraph data through zigzag persistence we use Reddit data with COVID-related subreddits. This data is known as the PAPERCRANE dataset [5].

The dataset subset we use spans from 1/20/20 to 3/31/20. This section captures the initial formation of the subreddits during the onset of COVID-19. The active subreddits related to COVID-19 in the dataset during this time are listed in Table 1 with summary statistics on the number of threads and authors.

Table 1. Subreddits related to covid from the PAPERCRANE datatset with number of threads and authors of each subreddit

Subreddit	Active Dates	Threads	Authors
CCP_virus	3/27 - 3/31	169	79
COVID19	2/15 - 3/31	8668	22020
COVID19positive	3/13 - 3/31	1462	6682
China_Flu	1/20 - 3/31	55466	62944
Coronavirus	1/20 - 3/31	153025	396427
CoronavirusCA	3/01 - 3/31	2930	5370
CoronavirusRecession	3/19 - 3/31	1574	6548
CoronavirusUK	2/15 - 3/31	8654	10230
CoronavirusUS	2/16 - 3/31	18867	29809
Covid2019	2/16 - 3/31	2437	1531
cvnews	1/25 - 3/31	4233	2181
nCoV	1/20 - 3/31	3949	1902

In this analysis we only use the nCoV subreddit due to its manageable size and interpretability. The temporal intervals for the edges are constructed from the author interaction information. We construct edge intervals based on the first and last times an author posted in each thread. These intervals are visualized in the top subfigure of Fig. 8.

We set the window size of 1 h with a shift of 15 min. This window size captures the necessary granularity to see changes in the dynamics of the subreddit. Applying this sliding window results in 6899 windows. The number of nodes and edges of each hypergraph snapshot is shown in Fig. 8. This initial data exploration shows that the size of the subreddit initially increases to a peak popularity at approximately two weeks into the subreddit or day 14. After this, the size steadily decreases. The edge intervals in the top subfigure of Fig. 8 shows that the majority of intervals are very short, while a few exhibit long intervals lasting as long as 38 days. This initial exploration does not capture how the shape of the underlying hypergraph snapshots is evolving.

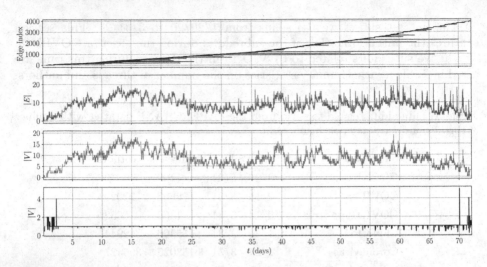

Fig. 8. Summary statistics for size of temporal hypergraph snapshots. The top is the interval associated to each edge (sorted by start time), the middle figure is the number of edges in the hypergraph snapshots, and the bottom figure is the number of vertices in the hypergraph snapshots.

There are many questions about the underlying network that can not be directly answered from these simple summary statistics. For example, is each thread dominated by unique authors or do many threads share users? Is the social network dense, centralized, fragmented? Do any of these characteristics change over time?

Understanding the topological summary of the hypergraph snapshots is important to understand the type of communication that is occurring. For example, many one-dimensional homology features are representative of disconnected conversations of holes in the communication structure. However, this could be captured just using the Betti sequence at each snapshot. What the zigzag persistence also captures is the severity of the holes based on their longevity. Consider a hole in communication that persists for several days. This could be representative of a lack of information communication throughout the community. These summary statistics additionally do not provide any information on how the threads in the subreddit are related and their longevity. Using zigzag persistence we can capture information about the longevity of communications using the zero-dimensional homology. A long interval in the zero-dimensional zigzag persistence barcode is representative of a conversation persisting over a long period of time. In Fig. 9 are the resulting zigzag persistence barcodes using the union between the associated ASCs of the hypergraph snapshots.

First, we see that we can capture how fragmented the social network is with one main component shown in the zero-dimensional barcode that persists for almost the entire duration of the subreddit. Additionally, the short intervals in dimension zero are characteristic of other side conversations, which either split from or merged into the main conversation or were entirely separate conversations. An example of one of these conversations is shown in the hypergraph snapshot at day 10 in Fig. 9 where the main component is composed of all of the threads with exception to one thread between just two authors. Having the main component suggests that many of the threads in the subreddit share at least one common author between threads.

We can also demonstrate that the network shows a change in its centralization over time. Specifically, during regions where many D_1 persistence intervals are present we know that the network has several loops, which are characteristic of non-centralized social networks. These changes from centralized to noncentralized social hypergraph snapshots are likely due to the number of authors active and a bifurcation of social network dynamics. For example, in the snapshot at day 10 in Fig. 9 there is a main loop in the main component of the hypergraph snapshot captured, and the main component does not have a clearly centralized structure. However, approximately one week later at day 18, there is a clearly centralized structure to the hypergraph which has no one-dimensional features. With both a low number of (or no) one-dimensional features and only one component, the zigzag persistence can give insight into the centralization of the hypergraph and underlying social network.

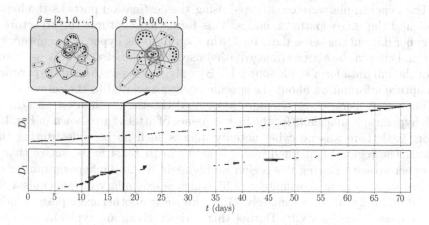

Fig. 9. Zigzag Persistence of temporal hypergraph representation of the CCP_virus subreddit with example hypergraph snapshot and associated ASC.

3.2 Cyber Data Analysis

For the analysis of cyber data we use the Operationally Transparent Cyber dataset (OpTC) [2] created by the Defense Advanced Research Projects Agency (DARPA). This dataset consists of network and host logging from hundreds of windows hosts over a week period. The dataset consists of two groups of user activity: benign and malicious. The malicious activity occurs over a three day period in which several attacks are executed.

Our goal is to analyze demonstrate how these attacks show up in the zigzag persistence barcodes for a hypergraph formation from the data log. The data log is composed of 64 columns describing each action in the network. In this section we only use the timestamps, src_ip, image_path, and dest_port, as these are needed to construct the temporal hypergraph representation of the data we study using zigzag persistence.

We construct hypergraph snapshots by again using a sliding window procedure, but now the intervals associated to each edge are only time points as the cyber data only has the time stamp at which the action occurred. We used a sliding window with width $w = 30$ min and shift $s = 5$ min. We chose this window size based on the duration of malicious activity lasting for approximately 2 h with 30 min windows being fine grained enough to capture the transition from benign to malicious.

To demonstrate how zigzag persistence can detect a cyber attack we will look at two instances of malicious activity on two different hosts. Namely, we investigate two cases of a cyber attack; the first on 9/23/19 from red agent LUAVR71T with source IP 142.20.56.202 on host 201 and the second on 9/24/19 from agent 4BW2MKUF with source IP 142.20.57.246 on host 501. The first sequence of attack beings at approximately 11:23 to 13:24 on 9/23/19 and the second sequence from approximately 10:46 to 13:11.

The hypergraphs were constructed using the destination ports as the hyperedges and the image paths as nodes. This formation captures the structure of the cyber data in the sense that the destination ports as hyperedges capture the relation between the actions (image paths) used. Additionally, we only use a subset of the full data for a single source IP. By only looking at this sub-hypergraph we capture information about the specific agent associated to the source IP.

The zigzag persistence barcodes associated the the destination port/image path hypergraph snapshots for the first sequence of attacks are shown in Fig. 10a. Before 9:00 there was no cyber activity and as such no barcodes during that period. The region highlighted in red from 11:23 to 13:24 is the active time of the cyber attacks. During this region we highlight a specific hypergraph for the window starting at approximately 12:35 which is exemplary of malicious activity. Additionally, at approximately 21:50, we show another exemplary window on standard benign activity. During this activity there are typically only two singletons which persist over time. A similar pair of hypergraphs for malicious and benign activity are shown in the second sequence of malicious activity on 9/24/19. However, what is not captured by the snapshots are the dynamics and quickly changing topology of the snapshots during malicious activity and relatively stationary dynamics and simple topology during benign activity.

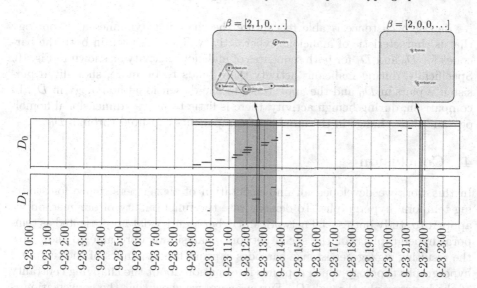

(a) Zigzag persistence barcodes of red team agent LUAVR71T with source IP 142.20.56.202 on host 201 for the day of 9/23/19.

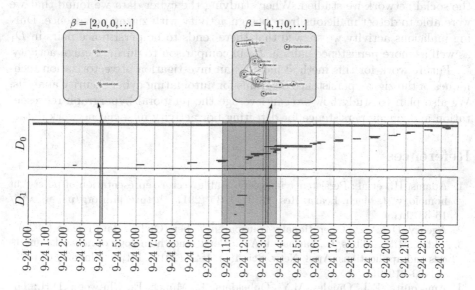

(b) Zigzag persistence barcodes of red team agent 4BW2MKUF with source IP 142.20.57.246 on host 501 for the day of 9/24/19.

Fig. 10. Zigzag persistence barcodes with example hypergraphs at two windows for OpTC data during an attack on the 23rd and 24th. The region highlighted in red is the time the red team agent was activity attacking. (Color figure online)

Zigzag persistence is able to capture the changing dynamics and topology that is characteristic of malicious cyber activity. This is shown in both the barcodes for D_0 and D_1 for both sequences of malicious activity as shown in Fig. 10. Specifically, during malicious activity there tends to be more, short-lived persistence pairs in D_0 and the appearance of one-dimensional homology in D_1. In comparison, during benign activity, there is little to no one-dimensional homology and little change in the number of components captured through D_0.

4   Conclusion

In this work we developed an implementation of zigzag persistence for studying temporal hypergraphs. To demonstrate the functionality of our method we apply it to study both social network and cyber security data represented as temporal hypergraphs. For the social network analysis we were able to show how the resulting zigzag persistence barcodes capture the dynamics of the temporal hypergraphs topology which captures information about the changing centrality of the hypergraphs through D_1. Furthermore, we show that the conversation is composed of one main component that persists over the entire time period of the social network we studied. When studying the cyber data we found that we were able to detect malicious from benign activity with zigzag persistence. During malicious activity we showed that there tends to be persistence pairs in D_1 as well as more persistence pairs in D_0 in comparison to during benign activity.

Future work for this method includes an investigation of vectorization techniques of the zigzag persistence diagrams for automating cyber security analysis. We also plan to study how we can leverage the temporal hypergraph representation and zigzag persistence for detecting bot activity in social network data.

References

1. Adams, H., et al.: Persistence images: a stable vector representation of persistent homology. J. Mach. Learn. Res. **18**(8), 1–35 (2017). http://jmlr.org/papers/v18/16-337.html
2. Agency, D.A.R.P.: Operationally transparent cyber (OpTC) data release (2020)
3. Aktas, M.E., Akbas, E., Fatmaoui, A.E.: Persistence homology of networks: methods and applications. Appl. Netw. Sci. **4**(1), 1–28 (2019). https://doi.org/10.1007/s41109-019-0179-3
4. Amézquita, E.J., Quigley, M.Y., Ophelders, T., Munch, E., Chitwood, D.H.: The shape of things to come: topological data analysis and biology, from molecules to organisms. Dev. Dyn. **249**(7), 816–833 (2020). https://doi.org/10.1002/dvdy.175
5. Baumgartner, J., Zannettou, S., Keegan, B., Squire, M., Blackburn, J.: The pushshift reddit dataset. PUSHSHIFT (2020). https://doi.org/10.5281/zenodo.3608135. Reddit-hazelnut prepared for the Social Network ProblemShop (Jan 24-Feb 4, 2022). Ottawa, Canada. Derivative of Reddit data obtained via pushshift.io API for the period January 1, 2019 to February 28
6. Bubenik, P.: Statistical topological data analysis using persistence landscapes. J. Mach. Learn. Res. **16**(3), 77–102 (2015). http://jmlr.org/papers/v16/bubenik15a.html

7. Carlsson, G., de Silva, V.: Zigzag persistence. Found. Comput. Math. **10**(4), 367–405 (2010). https://doi.org/10.1007/s10208-010-9066-0
8. Cencetti, G., Battiston, F., Lepri, B., Karsai, M.: Temporal properties of higher-order interactions in social networks. Sci. Rep. **11**(1) (2021). https://doi.org/10.1038/s41598-021-86469-8
9. David Boyce, B.R.: Modeling Dynamic Transportation Networks. Springer, Berlin Heidelberg (2012)
10. Edelsbrunner, L.: Zomorodian: topological persistence and simplification. Discrete Comput. Geom. **28**(4), 511–533 (2002). https://doi.org/10.1007/s00454-002-2885-2
11. Estrada, E., Rodríguez-Velázquez, J.A.: Subgraph centrality and clustering in complex hyper-networks. Phys. A: Stat. Mech. Appl. **364**, 581–594 (2006). https://doi.org/10.1016/j.physa.2005.12.002
12. Feng, S., et al.: Hypergraph models of biological networks to identify genes critical to pathogenic viral response. BMC Bioinf. **22**(1), 1–21 (2021). https://doi.org/10.1186/s12859-021-04197-2
13. Fischer, M.T., Arya, D., Streeb, D., Seebacher, D., Keim, D.A., Worring, M.: Visual analytics for temporal hypergraph model exploration. IEEE Trans. Vis. Comput. Graph. **27**(2), 550–560 (2021). https://doi.org/10.1109/tvcg.2020.3030408
14. Gasparovic, E., et al.: Homology of graphs and hypergraphs (2021). https://www.youtube.com/watch?v=XeNBysFcwOw
15. Golczynski, A., Emanuello, J.A.: End-to-end anomaly detection for identifying malicious cyber behavior through NLP-based log embeddings. arXiv preprint arXiv:2108.12276 (2021)
16. Hanselmann, M., Strauss, T., Dormann, K., Ulmer, H.: CANet: an unsupervised intrusion detection system for high dimensional can bus data. IEEE Access **8**, 58194–58205 (2020)
17. Harary, F., Gupta, G.: Dynamic graph models. Math. Comput. Model. **25**(7), 79–87 (1997). https://doi.org/10.1016/s0895-7177(97)00050-2
18. Husein, I., Mawengkang, H., Suwilo, S., Mardiningsih: modeling the transmission of infectious disease in a dynamic network. J. Phys.: Conf. Ser. **1255**(1), 012052 (2019). https://doi.org/10.1088/1742-6596/1255/1/012052
19. Joslyn, C.A., et al.: Hypernetwork science: from multidimensional networks to computational topology. In: Braha, D., et al. (eds.) ICCS 2020. SPC, pp. 377–392. Springer, Cham (2021). https://doi.org/10.1007/978-3-030-67318-5_25
20. Khasawneh, F., Munch, E.: Chatter detection in turning using persistent homology. Mech. Syst. Signal Process. **70–71**, 527–541 (2016). https://doi.org/10.1016/j.ymssp.2015.09.046
21. Munch, E.: A user's guide to topological data analysis. J. Learn. Anal. 4(2), 47–61 (2017). https://doi.org/10.18608/jla.2017.42.6
22. Myers, A., Munch, E., Khasawneh, F.A.: Persistent homology of complex networks for dynamic state detection. Phys. Rev. E **100**(2), 022314 (2019). https://doi.org/10.1103/physreve.100.022314
23. Myers, A., Muñoz, D., Khasawneh, F., Munch, E.: Temporal network analysis using zigzag persistence. EPJ Data Sci. **12**(1), 6 (2022)
24. Otter, N., Porter, M.A., Tillmann, U., Grindrod, P., Harrington, H.A.: A roadmap for the computation of persistent homology. EPJ Data Sci. **6**(1), 1–38 (2017). https://doi.org/10.1140/epjds/s13688-017-0109-5
25. Ren, S.: Persistent homology for hypergraphs and computational tools—a survey for users. J. Knot Theory Ramifications **29**(13), 2043007 (2020). https://doi.org/10.1142/s0218216520430075

26. Schäfer, B., Witthaut, D., Timme, M., Latora, V.: Dynamically induced cascading failures in power grids. Nat. Commun. **9**(1), 1975 (2018). https://doi.org/10.1038/s41467-018-04287-5
27. Skaf, Y., Laubenbacher, R.: Topological data analysis in biomedicine: a review. J. Biomed. Inf. **130**, 104082 (2022). https://doi.org/10.1016/j.jbi.2022.104082
28. Skyrms, B., Pemantle, R.: A dynamic model of social network formation. Proc. Natl. Acad. Sci. **97**(16), 9340–9346 (2000). https://doi.org/10.1073/pnas.97.16.9340
29. Tempelman, J.R., Khasawneh, F.A.: A look into chaos detection through topological data analysis. Phys. D: Nonlinear Phenom. **406**, 132446 (2020). https://doi.org/10.1016/j.physd.2020.132446
30. Tymochko, S., Munch, E., Khasawneh, F.: Using zigzag persistent homology to detect Hopf bifurcations in dynamical systems. Algorithms **13**(11), 278 (2020). https://doi.org/10.3390/a13110278
31. Xu, M., Radhakrishnan, S., Kamarthi, S., Jin, X.: Resiliency of mutualistic supplier-manufacturer networks. Sci. Rep. **9**(1), 1–10 (2019). https://doi.org/10.1038/s41598-019-49932-1
32. Yesilli, M.C., Chumley, M.M., Chen, J., Khasawneh, F.A., Guo, Y.: Exploring surface texture quantification in piezo vibration striking treatment (PVST) using topological measures. In: Volume 2: Manufacturing Processes; Manufacturing Systems. American Society of Mechanical Engineers (2022). https://doi.org/10.1115/msec2022-86659
33. Zomorodian, A., Carlsson, G.: Computing persistent homology. Discrete Comput. Geom. **33**(2), 249–274 (2004). https://doi.org/10.1007/s00454-004-1146-y

PageRank Nibble on the Sparse Directed Stochastic Block Model

Sayan Banerjee, Prabhanka Deka$^{(\boxtimes)}$, and Mariana Olvera-Cravioto

University of North Carolina, Chapel Hill, NC 27514, USA
{sayan,deka,molvera}@email.unc.edu

Abstract. We present new results on community recovery based on the PageRank Nibble algorithm on a sparse directed stochastic block model (dSBM). Our results are based on a characterization of the local weak limit of the dSBM and the limiting PageRank distribution. This characterization allows us to estimate the probability of misclassification for any given connection kernel and any given number of seeds (vertices whose community label is known). The fact that PageRank is a local algorithm that can be efficiently computed in both a distributed and asynchronous fashion, makes it an appealing method for identifying members of a given community in very large networks where the identity of some vertices is known.

Keywords: PageRank Nibble · directed stochastic block model · local weak convergence · community detection

1 Introduction

Many real-world networks exhibit community structure, where members of the same community are more likely to connect to each other than to members of different communities. Stochastic block models are frequently used to model random graphs with a community structure, and there are many problems where the goal is to identify the members of a given community, often based only on the graph structure, i.e., on the vertices and the existing edges among them. A two community symmetric SBM is described by two parameters α and β, which determine the edge probabilities, with α corresponding to the probability that two members from the same community connect to each other, and β to the probability that two members from different communities connect to each other. In [6], the authors work on the semi-sparse regime $\alpha = a \log n / n$ and $\beta = b \log n / n$, where n is the number of vertices in the graph, and show that the exact recovery of communities is efficiently possible if $|\sqrt{a} - \sqrt{b}| > 2$ and impossible otherwise. When recovery is possible, the authors use spectral methods to get an initial guess of the partition and fine tune it to retrieve the communities. Similar

Supported in part by the NSF RTG award (DMS-2134107) and in part by the NSF-CAREER award (DMS-2141621).

work has been done in the sparse regime, where $\alpha = a/n$ and $\beta = b/n$. In [7], the authors show that recovery is impossible when $(a-b)^2 < 2(a+b)$. In [8], it was proved that recovery is efficiently possible when $(a-b)^2 > 2(a+b)$ through the use of the spectral properties of a modified adjacency matrix B that counts the number of self avoiding paths of a given length l between two vertices in the graph. Further, the authors of [9] show that it is possible to recover a fraction $(1-\gamma)$ of the vertices of community 1 if a and b are sufficiently large and satisfy $(a-b)^2 > K_1 \log(\gamma^{-1})(a+b)$ for some constant K_1. The clustering methods in [6,8,9] all rely on finding eigenvectors of the adjacency matrix (or a modified adjacency matrix), which is computationally expensive for large networks.

Although the literature on community detection is vast, and there are in fact many methods that work remarkably well, many of those methods become computationally costly for very large networks. In some important cases like the web graph and social media networks, the networks of interest are so large and constantly changing that it becomes difficult to implement some of these methods. Moreover, in many cases, one has more information about the network than just its structure, e.g., vertex attributes that tell us the community to which certain vertices belong to. The question is then whether one can leverage knowledge of such vertices to help identify other members of their community using a computationally efficient method that does not require information about the entire network. One such problem was studied in [12], where the authors consider community detection in a dense (average degree of vertices scale linearly with the size of the network) SBM in which information about the presence or absence of each edge was hidden at random. Here, we will analyze a setting where the labels of some prominent members of the community of interest are known.

The PageRank Nibble algorithm was introduced in [11] as a modification of the Nibble algorithm described in [10] that uses personalized PageRank. This algorithm provides a cheap method for identifying the members of one community when a number of individuals in that community have been identified. PageRank based clustering methods were also proposed in [4] for the two-community SBM, as a special case of a more general method of combining random walk probabilities using a "discriminant" function.

The intuition behind PageRank Nibble is that random walks that start with the individuals that are known to belong to the community we seek will tend to visit more often members of that same community. PageRank Nibble works by choosing the personalization parameter of the known individuals, which we refer to as the "seeds", to be larger than for all other vertices in the network, and then choosing a damping factor c sufficiently far from either 0 or 1. This choice of the personalization values makes the PageRanks of close neighbors of the seeds to be larger, compared to those of individuals outside the community. Once the ranks produced by PageRank Nibble have been computed, a simple threshold rule can be used to identify the likely members of the community of interest. PageRank based methods can generally be executed quickly due to the availability of fast, distributed algorithms [13].

PageRank Nibble on the undirected SBM was studied in [1] under regimes where personalized PageRank (PPR) concentrates around its mean field approximation. The idea proposed there was to use the mean field approximation to identify vertices belonging to the same community as the seeds. In particular, the authors of [1] show that concentration occurs provided the average degrees grow as $a(n) \log n$ for some $a(n) \to \infty$ as $n \to \infty$, and is impossible for the sparse regime where average degrees remain constant as the network size grows. Our present work focuses on the directed stochastic block model (dSBM) in the sparse regime, and our results are based on the existence of a local weak limit and, therefore, of a limiting PageRank distribution. Once we have this characterization, we can compute the probability that an individual will be correctly or incorrectly classified, and choose the threshold that minimizes the misclassification probability.

2 Main Results

Let $\mathcal{G}_n = G(V_n, E_n)$ be a dSBM on the vertex set $V_n = \{1, \dots, n\}$ with two communities. To start, each vertex $v \in V_n$ is assigned a latent label $C_v \in \{1, 2\}$ identifying its community. We assume that these labels are unknown to us. Denote by $\mathcal{C}_1$ and $\mathcal{C}_2$ the subsets of vertices in communities 1 and 2 respectively. Then, each possible directed edge is sampled independently according to:

$$p_{vw}^{(n)} := \mathbb{P}((v, w) \in E_n | C_v, C_w) = \begin{cases} \frac{a}{n} \wedge 1 & \text{if } C_v = C_w \\ \frac{b}{n} \wedge 1 & \text{if } C_v \neq C_w. \end{cases}$$

The edge probabilities can be written as $p_{vw}^{(n)} = (n^{-1} \kappa_{C_v, C_w}) \wedge 1$, where

$$\kappa = \begin{bmatrix} a & b \\ b & a \end{bmatrix}$$

is called the connection probability kernel for the dSBM.

For $i = 1, 2$, we define

$$\pi_i^{(n)} = \frac{1}{n} \sum_{v=1}^{n} 1(C_v = i)$$

to be the proportion of vertices belonging to each community. We focus specifically on the case where $\pi_1^{(n)} = \pi_2^{(n)} = 1/2$, but the techniques used here can be applied to more general dSBMs.

To describe the setting for our results, start by fixing a constant $0 < s < 1$, and assume there exists a subset $\mathcal{S} \subseteq \mathcal{C}_1$, with $|\mathcal{S}| = n\pi_1^{(n)} s$, for which the community labels are known. In other words, we assume that we know the identities of a fixed, positive proportion of the vertices in community 1. We refer to the vertices in $\mathcal{S}$ as the *seeds*. In a real-world social network one can think of the seeds as famous individuals whose community label or affiliation is known or easy to infer. Given the seed set $\mathcal{S}$, the goal is to identify the vertices $v \in \mathcal{C}_1 \setminus \mathcal{S}$, i.e., to recover the remaining members of community 1.

In order to describe the PageRank Nibble algorithm, we start first with the definition of personalized PageRank. On a directed graph $G = (V, E)$, the PageRank of vertex $v \in V$ is given by:

$$r_v = c \sum_{w \in V:(w,v) \in E} \frac{1}{D_w^+} r_w + (1 - c)q_v, \tag{1}$$

where D_w^+ is the out-degree of vertex $w \in V$, q_v is the personalization value of vertex v, and $c \in (0, 1)$ is a damping factor.

PageRank is one of most popular measures of network centrality, due to both its computational efficiency (it can be computed in a distributed and asynchronous way), and its ability to identify *relevant* vertices. When $\mathbf{q} = (q_v : v \in V)$ is a probability vector, the PageRank vector $\mathbf{r} = (r_v : v \in V)$ is known to correspond to the stationary distribution of a random walk that, at each time step, chooses, with probability c, to follow an outbound edge uniformly chosen at random, or with probability $1 - c$, chooses its next destination according to $\mathbf{q}$ (if the current vertex has no outbound edges, the random walk always chooses its next destination according to $\mathbf{q}$). PageRank is known to rank highly vertices that either have a large in-degree, or that have close inbound neighbors whose PageRanks are very large [14], hence capturing both popularity and credibility. Since on large networks the PageRank scores will tend to be very small, it is often convenient to work with the scale-free (graph-normalized) PageRank vector $\mathbf{R} = |V|\mathbf{r}$ instead.

For the two community dSBM $G_n = (V_n, E_n)$ described above, let $Q_v = nq_v$ and define

$$\mu_n(B) = \frac{1}{n} \sum_{v=1}^{n} 1((C_v, Q_v) \in B)$$

for any measurable set B. We assume that there exists a limiting measure μ with $\pi_i := \mu(\{i\} \times \mathbb{R}_+) > 0$ for $i = 1, 2$ such that

$$\mu_n \Rightarrow \mu \tag{2}$$

in probability. Here and in the sequel, $\Rightarrow$ denotes weak convergence. Further, for any measurable A, let

$$\sigma_i^{(n)}(A) = \frac{1}{n\pi_i^{(n)}} \sum_{v \in V_n} 1(C_v = i, Q_v \in A), \qquad i = 1, 2, \tag{3}$$

denote the empirical distribution of Q_v conditionally on $C_v = i$ for $i = 1, 2$. Due to assumption (2), we get the existence of limiting distributions σ_i, given by

$$\sigma_i(A) = \frac{\mu(\{i\} \times A)}{\pi_i}, \qquad i = 1, 2,$$

such that $\sigma_i^{(n)} \Rightarrow \sigma_i$ in probability as $n \to \infty$.

As mentioned in the introduction, our analysis is based on the existence of a local weak limit for the dSBM, and the fact that if we let I be uniformly chosen in V_n, independently of $G(V_n, E_n)$, and let R_I denote the scale-free PageRank of vertex I, then R_I converges weakly to a random variable $\mathcal{R}$ as $n \to \infty$. In order to characterize the distribution of $\mathcal{R}$, first define $R^{(1)}$ and $R^{(2)}$ to be random variables satisfying

$$\mathbb{P}\left(R^{(i)} \in \cdot\right) = \mathbb{P}\left(R_I \in \cdot \,|\, C_I = i\right), \qquad i = 1, 2.$$

Our first result establishes the weak convergence of $R^{(i)}$ for $i = 1, 2$ and characterizes the limiting distributions as the solutions to a system of distributional fixed-point equations.

Theorem 1. *Let $G_n = (V_n, E_n)$ be a sequence of dSBM as described above such that (2) holds. Then, there exist random variables $\mathcal{R}^{(1)}$ and $\mathcal{R}^{(2)}$ such that for any $x \in \mathbb{R}$ that is a point of continuity of the limit,*

$$R^{(i)} \Rightarrow \mathcal{R}^{(i)} \qquad and \qquad \frac{2}{n} \sum_{v \in V_n} 1(R_v \leq x, C_v = i) \xrightarrow{P} \mathbb{P}\left(\mathcal{R}^{(i)} \leq x\right),$$

as $n \to \infty$, $i = 1, 2$. Moreover, the random variables $\mathcal{R}^{(1)}$ and $\mathcal{R}^{(2)}$ satisfy:

$$\mathcal{R}^{(1)} \stackrel{d}{=} c \sum_{j=1}^{\mathcal{N}^{(11)}} \frac{\mathcal{R}_j^{(1)}}{\mathcal{D}_j^{(1)}} + c \sum_{j=1}^{\mathcal{N}^{(12)}} \frac{\mathcal{R}_j^{(2)}}{\mathcal{D}_j^{(2)}} + (1-c)\mathcal{Q}^{(1)} \tag{4}$$

$$\mathcal{R}^{(2)} \stackrel{d}{=} c \sum_{j=1}^{\mathcal{N}^{(21)}} \frac{\mathcal{R}_j^{(1)}}{\mathcal{D}_j^{(1)}} + c \sum_{j=1}^{\mathcal{N}^{(22)}} \frac{\mathcal{R}_j^{(2)}}{\mathcal{D}_j^{(2)}} + (1-c)\mathcal{Q}^{(2)} \tag{5}$$

where $\mathcal{Q}^{(1)}$ and $\mathcal{Q}^{(2)}$ are random variables distributed according to σ_1 and σ_2 respectively, $\mathcal{N}^{(kl)}$ are Poisson random variables with means $\pi_l \kappa_{lk}$, $(\mathcal{D}_j^{(i)} - 1 : j \geq 1)$, $i = 1, 2$, are i.i.d. sequences of Poisson random variables with mean $\pi_1 \kappa_{i1} + \pi_2 \kappa_{i2}$, and $(\mathcal{R}_j^{(i)} : j \geq 1)$, $i = 1, 2$, are i.i.d. copies of $\mathcal{R}^{(i)}$, with all random variables independent of each other.

Remark 1. Note that the $(\mathcal{D}_j^{(i)})$ are size-biased Poisson random variables that represent the out-degrees of the inbound neighbors of the explored vertex I.

The above result holds in more generality for a degree-corrected dSBM with k-communities, but for the purposes of this paper, we restrict ourselves to the $k = 2$ case. We will only outline a sketch of the proof, and focus our attention instead on the following theorem about the classification of the vertices.

Equations (4) and (5) are the key behind our classification method. Observe that in the PageRank Eq. (1), the parameters within our control are the damping factor c and the personalization vector $\mathbf{Q} = (Q_v : v \in V_n)$. If we choose $\mathbf{Q}$ that results in $\mathcal{R}^{(1)} \geq_{\text{s.t.}} \mathcal{R}^{(2)}$, we can identify vertices in community 1 as the ones having higher PageRank scores. With that in mind, we set $Q_v = 1(v \in \mathcal{S})$, choose an appropriate cutoff point x_0 (which might depend on c, s and κ), and classify as a member of community 1 any vertex $v \in V_n$ such that its scale-free PageRank, R_v, satisfies $R_v > x_0$. The algorithm requires that we choose c sufficiently bounded away from both zero and one, since from the random walk interpretation of PageRank, it is clear that we want to give the random surfer time to explore the local neighborhood, while at the same time ensuring that it returns sufficiently often to the seed set. In practice, a popular choice for the damping factor is $c = 0.85$. In the context of the dSBM, we have that when $a >> b$, the random surfer ends up spending more time exploring the vertices in community 1, and the probability that it escapes to community 2 before jumping back to the seeds is much smaller. As a result, the stationary distribution ends up putting more mass on the community 1 vertices, and the proportion of misclassified vertices diminishes when $a + b$ is large and b/a is close to zero. We formalize this in the theorem below. Note that Theorem 1 gives that the misclassification probabilities satisfy:

$$\mathbb{P}\left(R_v \le x_0 | v \in \mathcal{C}_1 \right) \approx \mathbb{P}\left(\mathcal{R}^{(1)} \le x_0 \right) \quad \text{and}$$

$$\mathbb{P}\left(R_v > x_0 | v \in \mathcal{C}_2 \right) \approx \mathbb{P}\left(\mathcal{R}^{(2)} > x_0 \right).$$

Our local classification algorithm with input parameters c and x_0 is then described as follows:

1. Set $Q_v = 1$ if $v \in \mathcal{S}$, and zero otherwise.
2. Fix the damping factor $c \in (0, 1)$ and compute the personalized scale-free PageRank vector $\mathbf{R}$.
3. For a threshold x_0, the estimated members of $\mathcal{C}_1$ are the vertices in the set $\hat{\mathcal{C}}_1(x_0, c) = \{v \in V_n : R_v > x_0\}$.

The theorem below can be used to quantify the damping factor c and the classification threshold x_0, and the corollary that follows shows that the proportion of misclassified vertices becomes small with high probability as $n \to \infty$.

Theorem 2. *Let $G_n = (V_n, E_n)$ be a 2-community dSBM with*

$$\kappa = \begin{bmatrix} a & b \\ b & a \end{bmatrix}$$

and $\pi_1 = \pi_2 = 1/2$. Assume a, b satisfy $8b/(a + b) < 1/2$ and $\mathrm{e}^{-(a+b)/2} < b/4a$. Let $Q_v = 1(v \in \mathcal{S})$ for $v \in V_n$, and take any $c \in (1/2, 1 - 8b/(a + b)]$. Then, for $x_0 = 5s/8$, we have

$$\mathbb{P}\left(\mathcal{R}^{(1)} < \frac{5s}{8} \right) \le \frac{256c^2}{(a+b)(1-c^2)} + \frac{64(1-c)(1-s)}{(1+c)s}, \tag{6}$$

$$\mathbb{P}\left(\mathcal{R}^{(2)} > \frac{5s}{8} \right) \le \frac{256c^2}{(a+b)(1-c^2)} \left(1 + \frac{(1-c)(1-s)}{2(1+c)s} \right). \tag{7}$$

Naturally, the misclassification errors get smaller as s increases, i.e., as more members of community 1 are known. Also, we get better bounds for the misclassification errors when $a + b$ is large (strong connectivity within a community) and $b/(a+b)$ is small (equivalently, $a/(a+b)$ close to one), i.e., when the network is strongly assortative.

Note that the assumptions in Theorem 2 do not involve s (proportion of seeds). As the proof indicates, our classification errors involve Chebychev bounds which crucially depend on: (i) the mean PageRank scores of the two communities being sufficiently different, and (ii) the ratio of the variance of the PageRank scores of vertices in each community to the square of the mean community PageRank being small. By Lemma 1 below, the ratio of the mean community PageRank scores is independent of s and hence their separation required by (i) is ensured by conditions involving a, b but not s. Moreover, as seen in Lemma 2, the scaled fluctuations in (ii) depend more significantly on the 'sparsity' of the underlying network, quantified by $a + b$ (expected total degree of a vertex), than s. Thus, the dependence on s arises mainly through the choice of the threshold x_0 in our classification algorithm (see Corollary 1).

As a direct corollary to Theorem 2, we have

Corollary 1. *Let* $x_0 = 5s/8$, $c \in (1/2, 1 - 8b/(a + b)]$,

$$\delta_1 = \frac{256c^2}{(a+b)(1-c^2)} + \frac{64(1-c)(1-s)}{(1+c)s}$$

and

$$\delta_2 = \frac{256c^2}{(a+b)(1-c^2)}\left(1 + \frac{(1-c)(1-s)}{2(1+c)s}\right).$$

Then, under the hypothesis of Theorem 2, for $\delta = \delta_1 + \delta_2$ *and any* $\epsilon > 0$, *we have*

$$\lim_{n \to \infty} \mathbb{P}\left(|\mathcal{C}_1 \triangle \hat{\mathcal{C}}_1(x_0, c)| > \frac{(\delta + \epsilon)n}{2}\right) = 0.$$

Proof. For notational convenience, we drop the dependence of $\hat{\mathcal{C}}_1$ on x_0 and c. Observe that $|\mathcal{C}_1 \triangle \hat{\mathcal{C}}_1| = |\mathcal{C}_1 \backslash \hat{\mathcal{C}}_1| + |\hat{\mathcal{C}}_1 \cap \mathcal{C}_2|$, and we have $\mathcal{C}_1 \backslash \hat{\mathcal{C}}_1 = \{v \in \mathcal{C}_1 : R_v < 5s/8\}$ and $\hat{\mathcal{C}}_1 \cap \mathcal{C}_2 = \{v \in \mathcal{C}_2 : R_v > 5s/8\}$. So we get that for $x_0 = 5s/8$,

$$\mathbb{P}\left(|\mathcal{C}_1 \triangle \hat{\mathcal{C}}_1| > \frac{(\delta + \epsilon)n}{2}\right) = \mathbb{P}\left(\frac{2}{n}\sum_{v \in \mathcal{C}_1} 1(R_v < x_0) + \frac{2}{n}\sum_{v \in \mathcal{C}_2} 1(R_v > x_0) > \delta + \epsilon\right).$$

Then the result follows since

$$\frac{2}{n}\sum_{v \in \mathcal{C}_1} 1(R_v < x_0) + \frac{2}{n}\sum_{v \in \mathcal{C}_2} 1(R_v > x_0)$$

$$\xrightarrow{P} \mathbb{P}(\mathcal{R}^{(1)} < x_0) + \mathbb{P}(\mathcal{R}^{(2)} > x_0) = \delta$$

as $n \to \infty$.

Remark 2. Our proof of Theorem 2 uses Chebyshev's inequalities based on mean and variance bounds for the limiting (scale-free) personalized PageRank distribution obtained from the distributional fixed-point equations in Theorem 1. The choice of x_0 above is rather ad hoc and mainly for simplicity of the associated misclassification error bounds. One can check that the choice of x_0 which minimizes the sum of the Chebyshev error bounds is given by $x_0^* = (r_1 v_2^{1/3} + r_2 v_1^{1/3})/(v_1^{1/3} + v_2^{1/3})$, where r_1, r_2 are the expected limiting PageRank values obtained in Lemma 1 and v_1, v_2 are the corresponding variances obtained in Lemma 2. Further, $x_0 = 5s/8$ is independent of the kernel parameters a and b, which are often unknown in practice. Moreover, although the range of c depends on a, b, the results above hold for any c in the given range. Hence, in practice, when a, b are not known, then any $c > 1/2$ which is not too close to one should work provided the network is not too sparse ($b/(a + b)$ is sufficiently small).

3   Proofs

As mentioned earlier, Theorem 1 holds in considerably more generality than the one stated here, so we will only provide a sketch of the proof that suffices for the simpler dSBM considered here. The proof of Theorem 2 is given later in the section.

Proof. Theorem 1 (Sketch). The proof consists of three main steps.

1. **Establish the local weak convergence of the dSBM:** For the 2-community dSBM considered here, one can modify the coupling in [3] (which works for an undirected SBM) to the exploration of the in-component of a uniformly chosen vertex. The coupled graph is a 2-type Galton-Watson process, with the two types corresponding to the two communities in the dSBM, and all edges directed from offspring to parent. The number of offspring of type j that a node of type i has is a Poisson random variable with mean $m_{ij}^- = \pi_j \kappa_{ji}$ for $j = 1, 2$. For each node $\mathbf{i}$ in the coupled tree, denote by $C_{\mathbf{i}}$ its type, and assign it a mark $X_{\mathbf{i}} = (\mathcal{D}_{\mathbf{i}}, Q_{\mathbf{i}})$, where $(\mathcal{D}_{\mathbf{i}} - 1)|C_{\mathbf{i}} = j$ is a Poisson random variable with mean $m_j^+ = \pi_1 \kappa_{j1} + \pi_2 \kappa_{j2}$, and $Q_{\mathbf{i}}|C_{\mathbf{i}} = j$ has distribution σ_j as defined in (3). The construction of the coupling follows a two step exploration process similar to the one done for inhomogeneous random digraphs in [5]. First the outbound edges of a vertex are explored, followed by the exploration of its inbound neighbors, assigning marks to a vertex once we finish exploring both its inbound and outbound one-step neighbors. This establishes the local weak convergence in probability of the dSBM to the 2-type Galton-Watson process.

2. **Establish the local weak convergence of PageRank:** Once we have the local weak convergence of the dSBM, let $\mathcal{R}^*$ denote the personalized PageRank of the root node of the 2-type Galton-Watson process in the coupling. The local weak convergence in probability of the PageRanks on the dSBM to $\mathcal{R}^*$, i.e.,

$$\frac{1}{n} \sum_{v \in V_n} 1(R_v \le x) \xrightarrow{P} \mathbb{P}(\mathcal{R}^* \le x)$$

as $n \to \infty$, follows from Theorem 2.1 in [2]. Note that the random variables $\mathcal{R}^{(1)}$ and $\mathcal{R}^{(2)}$ correspond to the conditional laws of $\mathcal{R}^*$ given that the root has type 1 or type 2, respectively. And since the two communities are assumed to have the same size, the probability that the root has type 1 is $1/2$, hence,

$$\frac{1}{n} \sum_{v \in V_n} 1(R_v \le x, C_v = i) \xrightarrow{P} \mathbb{P}(\mathcal{R}^{(i)} \le x)\frac{1}{2},$$

as $n \to \infty$. The weak convergence result follows from the bounded convergence theorem.

3. **Derive the distributional fixed point equations:** If the nodes in the first generation of the 2-type Galton-Watson process are labeled $1 \le j \le \mathcal{N}$, where $\mathcal{N}$ denotes the number of offspring of the root node, then

$$\mathcal{R}^* = c \sum_{j=1}^{\mathcal{N}} \frac{\mathcal{R}_j}{\mathcal{D}_j} + (1-c)\mathcal{Q},$$

where $\mathcal{Q}$ denotes the personalization value of the root, $(\mathcal{D}_j : j \ge 1)$ correspond to the out-degrees of the offspring, and the $(\mathcal{R}_j : j \ge 1)$ correspond to their PageRanks. Conditioning on the type of the root, as well as on the types of its offspring, gives the two distributional fixed-point Eqs. (4) and (5). In particular, conditionally on the root having type i, $\mathcal{N}^{(ik)}$ corresponds to the number of offspring of type k, $\mathcal{Q}^{(i)}$ has distribution σ_i, and $\mathcal{D}_1^{(k)}$ and $\mathcal{R}_1^{(k)}$ are independent random variables having the distribution of $\mathcal{D}_1$ and $\mathcal{R}_1$ conditionally on node 1 having type k.

We prove Theorem 2 through the second moment method. First we prove the following lemmas establishing bounds on the mean and variance of $\mathcal{R}^{(1)}$ and $\mathcal{R}^{(2)}$.

Lemma 1. *Let* $r_i = \mathbb{E}\left[\mathcal{R}^{(i)}\right]$, $\lambda = 1 - e^{-(a+b)/2}$ *and*

$$\Delta = \left(1 - \frac{c\lambda a}{a+b}\right)^2 - \left(\frac{c\lambda b}{a+b}\right)^2.$$

Then, we have

$$r_1 = \frac{\left(1 - \frac{c\lambda a}{a+b}\right)s(1-c)}{\Delta} \tag{8}$$

$$r_2 = \frac{\left(\frac{c\lambda b}{a+b}\right)s(1-c)}{\Delta}. \tag{9}$$

Further, if $1 - \lambda = e^{-(a+b)/2} \le b/4a$ *and* $c > 1/2$, *we have the bounds*

$$r_1 \ge s\left(1 - \frac{2b}{(1-c)(a+b)}\right), \tag{10}$$

$$r_2 \le \frac{s}{2}. \tag{11}$$

Proof. Recall the distributional equations satisfied by $\mathcal{R}^{(1)}$ and $\mathcal{R}^{(2)}$ from Theorem 1. Taking expectation on both sides gives us

$$\mathbb{E}[\mathcal{R}^{(1)}] = c\mathbb{E}\left[\sum_{j=1}^{\mathcal{N}^{(11)}} \frac{\mathcal{R}_j^{(1)}}{\mathcal{D}_j^{(1)}} + \sum_{j=1}^{\mathcal{N}^{(12)}} \frac{\mathcal{R}_j^{(2)}}{\mathcal{D}_j^{(2)}}\right] + (1-c)\mathbb{E}[\mathcal{Q}^{(1)}],$$

$$\mathbb{E}[\mathcal{R}^{(2)}] = c\mathbb{E}\left[\sum_{j=1}^{\mathcal{N}^{(21)}} \frac{\mathcal{R}_j^{(1)}}{\mathcal{D}_j^{(1)}} + \sum_{j=1}^{\mathcal{N}^{(22)}} \frac{\mathcal{R}_j^{(2)}}{\mathcal{D}_j^{(2)}}\right] + (1-c)\mathbb{E}[\mathcal{Q}^{(2)}].$$

First, note that with our choice of $\mathcal{Q}$, $\mathbb{E}[\mathcal{Q}^{(1)}] = s$ and $\mathbb{E}[\mathcal{Q}^{(2)}] = 0$. Further $(\mathcal{R}_j^{(i)}, \mathcal{D}_j^{(i)})_{j\geq 1}$ (resp. $(\mathcal{R}_j^{(i)}, \mathcal{D}_j^{(i)})_{j\geq 1}$) are independent of $\mathcal{N}^{(1i)}$ (resp. $\mathcal{N}^{(2i)}$), and of each other, for $i = 1, 2$. So the above expressions can be simplified to

$$r_1 = c\left(\mathbb{E}[\mathcal{N}^{(11)}]\mathbb{E}\left[\frac{1}{\mathcal{D}^{(1)}}\right]r_1 + \mathbb{E}[\mathcal{N}^{(12)}]\mathbb{E}\left[\frac{1}{\mathcal{D}^{(2)}}\right]r_2\right) + (1-c)s,$$

$$r_2 = c\left(\mathbb{E}[\mathcal{N}^{(21)}]\mathbb{E}\left[\frac{1}{\mathcal{D}^{(1)}}\right]r_1 + \mathbb{E}[\mathcal{N}^{(22)}]\mathbb{E}\left[\frac{1}{\mathcal{D}^{(2)}}\right]r_2\right),$$

where $\mathcal{N}^{(ij)}$ and $(\mathcal{D}^{(i)} - 1)$ are Poisson random variables with means as described in Theorem 1. Therefore we can further reduce the equations to

$$r_1 = c\left(\frac{a}{2}\cdot\frac{(1-e^{-(a+b)/2})}{(a+b)/2}\cdot r_1 + \frac{b}{2}\cdot\frac{(1-e^{-(a+b)/2})}{(a+b)/2}\cdot r_2\right) + (1-c)s,$$

$$r_2 = c\left(\frac{b}{2}\cdot\frac{(1-e^{-(a+b)/2})}{(a+b)/2}\cdot r_1 + \frac{a}{2}\cdot\frac{(1-e^{-(a+b)/2})}{(a+b)/2}\cdot r_2\right),$$

or in matrix form, and after substituting $\lambda = (1 - e^{-(a+b)/2})$,

$$\begin{bmatrix} 1 - \frac{ca\lambda}{a+b} & -\frac{cb\lambda}{a+b} \\ -\frac{cb\lambda}{a+b} & 1 - \frac{ca\lambda}{a+b} \end{bmatrix}\begin{bmatrix} r_1 \\ r_2 \end{bmatrix} = \begin{bmatrix} (1-c)s \\ 0 \end{bmatrix}. \tag{12}$$

Solving (12), we get

$$\begin{bmatrix} r_1 \\ r_2 \end{bmatrix} = \frac{1}{\Delta}\begin{bmatrix} (1 - c\lambda a/(a+b))\, s(1-c) \\ c\lambda bs(1-c)/(a+b) \end{bmatrix},$$

where

$$\Delta = \left(1 - \frac{c\lambda a}{a+b}\right)^2 - \left(\frac{c\lambda b}{a+b}\right)^2$$

as required. Note that since $c\lambda < 1$, we have $\Delta > 0$, and so the above quantities are well defined. From here, the bound for r_2 is a straightforward calculation.

$$r_2 = \frac{\frac{c\lambda b}{a+b}(1-c)s}{\Delta} \leq \frac{\frac{b}{a+b}(1-c)s}{(1-c\lambda)\left(1-c\lambda\frac{a-b}{a+b}\right)} \leq \frac{sb}{(a+b)}\frac{1}{\left(1-\frac{a-b}{a+b}\right)} = \frac{s}{2}.$$

To get the bound for r_1, we proceed as follows

$$
\begin{aligned}
r_1 &= \frac{\left(1 - \frac{c\lambda a}{a+b}\right)(1-c)s}{\Delta} \\
&\geq \frac{s(1-c)}{1 - \frac{c\lambda a}{a+b}} = \frac{s(1-c)}{\frac{a+b}{a+b} - \frac{c\lambda a}{a+b}} = \frac{s(1-c)}{\frac{b}{a+b} + \frac{a}{a+b}(1 - \lambda + \lambda(1-c))} \\
&\geq \frac{s(1-c)}{\frac{b}{a+b} + \frac{a}{a+b}\left(e^{-(a+b)/2} + (1-c)\right)} = \frac{s(1-c)}{(1-c) + \frac{cb}{a+b} + \frac{a}{a+b}e^{-(a+b)/2}} \\
&\geq \frac{s(1-c)}{1 - c + \frac{2bc}{a+b}} \geq s\left(1 - \frac{2b}{(1-c)(a+b)}\right),
\end{aligned}
$$

where for the last inequality we used fact that $e^{-(a+b)/2}a/(a+b) \leq b/4(a+b) \leq cb/(a+b)$ due to our assumptions on λ and c, and $1 - x^2 \leq 1$ for all $x \in \mathbb{R}$. This completes the proof.

The next lemma provides a result for the variances.

Lemma 2. *Define* $v_i = \mathrm{Var}(\mathcal{R}^{(i)})$ *for* $i = 1, 2$, *then, if we let* $\mathbf{v} = (v_1, v_2)'$, *and* $\mathbf{r}^2 = (r_1^2, r_2^2)'$, *then*

$$
\mathbf{v} = \frac{1}{2(1 - g_1)(1 - g_2)}\left(K\mathbf{r}^2 + (1-c)^2 s(1-s)\mathbf{k}\right),
$$

where $g_1 = c^2\mathbb{E}[1/(\mathcal{D}^{(1)})^2](a-b)/2$, $g_2 = c^2\mathbb{E}[1/(\mathcal{D}^{(1)})^2](a+b)/2$,

$$
K = \begin{bmatrix} g_1 + g_2 - 2g_1 g_2, & g_2 - g_1 \\ g_2 - g_1, & g_1 + g_2 - 2g_1 g_2 \end{bmatrix}, \quad \text{and} \quad \mathbf{k} = \begin{bmatrix} 2 - g_1 - g_2 \\ g_2 - g_1 \end{bmatrix}.
$$

Furthermore,

$$
v_1 \leq \frac{4c^2 s^2}{(a+b)(1-c^2)} + \frac{1-c}{1+c}s(1-s),
$$

$$
v_2 \leq \frac{4c^2 s^2}{(a+b)(1-c^2)}\left(1 + \frac{(1-c)(1-s)}{2s(1+c)}\right).
$$

Proof. To calculate the variance of $\mathcal{R}^{(1)}$ and $\mathcal{R}^{(2)}$, we will rely on the law of total variances, i.e., for any two random variables X and Y,

$$
\mathrm{Var}(X) = \mathrm{Var}[\mathbb{E}(X|Y)] + \mathbb{E}[\mathrm{Var}(X|Y)].
$$

Applying this to Eq. (4), along with the fact that $r_i < 1$ for $i = 1, 2$, we get the following bound for $\mathrm{Var}(\mathcal{R}^{(1)})$:

$$
\begin{aligned}
\mathrm{Var}(\mathcal{R}^{(1)}) &= c^2\mathrm{Var}\left(r_1\mathcal{N}^{(11)}\mathbb{E}\left[\frac{1}{\mathcal{D}^{(1)}}\right] + r_2\mathcal{N}^{(12)}\mathbb{E}\left[\frac{1}{\mathcal{D}^{(2)}}\right]\right) \\
&+ c^2\mathbb{E}\left[\mathcal{N}^{(11)}\mathrm{Var}\left(\frac{\mathcal{R}^{(1)}}{\mathcal{D}^{(1)}}\right) + \mathcal{N}^{(12)}\mathrm{Var}\left(\frac{\mathcal{R}^{(2)}}{\mathcal{D}^{(2)}}\right)\right] + (1-c)^2\mathrm{Var}(\mathcal{Q}^{(1)}).
\end{aligned}
$$

Now use the observation that $\mathrm{Var}(\mathcal{N}^{(11)}) = \mathbb{E}[\mathcal{N}^{(11)}] = a/2$, $\mathrm{Var}(\mathcal{N}^{(12)}) = \mathbb{E}[\mathcal{N}^{(12)}] = b/2$, and $\mathrm{Var}(\mathcal{Q}^{(1)}) = s(1-s)$, to obtain that for $v_i = \mathrm{Var}(\mathcal{R}^{(i)})$,

$$v_1 = c^2 \left(r_1^2 \cdot \frac{a}{2} \cdot \left(\mathbb{E}\left[\frac{1}{\mathcal{D}^{(1)}}\right]\right)^2 + r_2^2 \cdot \frac{b}{2} \cdot \left(\mathbb{E}\left[\frac{1}{\mathcal{D}^{(2)}}\right]\right)^2 \right)$$

$$+ c^2 \left(\frac{a}{2} \mathrm{Var}\left(\frac{\mathcal{R}^{(1)}}{\mathcal{D}^{(1)}}\right) + \frac{b}{2} \mathrm{Var}\left(\frac{\mathcal{R}^{(2)}}{\mathcal{D}^{(2)}}\right) \right) + (1-c)^2 s(1-s)$$

$$= c^2 \left(r_1^2 \cdot \frac{a}{2} \cdot \left(\mathbb{E}\left[\frac{1}{\mathcal{D}^{(1)}}\right]\right)^2 + r_2^2 \cdot \frac{b}{2} \cdot \left(\mathbb{E}\left[\frac{1}{\mathcal{D}^{(2)}}\right]\right)^2 \right) + (1-c)^2 s(1-s)$$

$$+ \frac{c^2 a}{2} \left(\mathrm{Var}\left(\frac{1}{\mathcal{D}^{(1)}} r_1\right) + \mathbb{E}\left[\frac{1}{(\mathcal{D}^{(1)})^2} \mathrm{Var}(\mathcal{R}^{(1)})\right] \right)$$

$$+ \frac{c^2 b}{2} \left(\mathrm{Var}\left(\frac{1}{\mathcal{D}^{(2)}} r_2\right) + \mathbb{E}\left[\frac{1}{(\mathcal{D}^{(2)})^2} \mathrm{Var}(\mathcal{R}^{(2)})\right] \right)$$

$$= c^2 \left(r_1^2 \cdot \frac{a}{2} \cdot \left(\mathbb{E}\left[\frac{1}{\mathcal{D}^{(1)}}\right]\right)^2 + r_2^2 \cdot \frac{b}{2} \cdot \left(\mathbb{E}\left[\frac{1}{\mathcal{D}^{(2)}}\right]\right)^2 \right) + (1-c)^2 s(1-s)$$

$$+ \frac{c^2 a}{2} \left(r_1^2 \mathrm{Var}\left(\frac{1}{\mathcal{D}^{(1)}}\right) + \mathbb{E}\left[\frac{1}{(\mathcal{D}^{(1)})^2}\right] v_1 \right)$$

$$+ \frac{c^2 b}{2} \left(r_2^2 \mathrm{Var}\left(\frac{1}{\mathcal{D}^{(2)}}\right) + \mathbb{E}\left[\frac{1}{(\mathcal{D}^{(2)})^2}\right] v_2 \right)$$

$$= (1-c)^2 s(1-s) + \frac{c^2 a}{2}\mathbb{E}\left[\frac{1}{(\mathcal{D}^{(1)})^2}\right](r_1^2 + v_1) + \frac{c^2 b}{2}\mathbb{E}\left[\frac{1}{(\mathcal{D}^{(2)})^2}\right](r_2^2 + v_2).$$

Similarly, using $\mathcal{Q}^{(2)} \equiv 0$ and

$$v_2 = c^2 \mathrm{Var}\left(r_1 \mathcal{N}^{(21)}\mathbb{E}\left[\frac{1}{\mathcal{D}^{(1)}}\right] + r_2 \mathcal{N}^{(22)}\mathbb{E}\left[\frac{1}{\mathcal{D}^{(2)}}\right] \right)$$

$$+ c^2 \mathbb{E}\left[\mathcal{N}^{(21)}\mathrm{Var}\left(\frac{\mathcal{R}^{(1)}}{\mathcal{D}^{(1)}}\right) + \mathcal{N}^{(22)}\mathrm{Var}\left(\frac{\mathcal{R}^{(2)}}{\mathcal{D}^{(2)}}\right)\right] + (1-c)^2 \mathrm{Var}(\mathcal{Q}^{(2)})$$

$$= \frac{c^2 b}{2}\mathbb{E}\left[\frac{1}{(\mathcal{D}^{(1)})^2}\right](r_1^2 + v_1) + \frac{c^2 a}{2}\mathbb{E}\left[\frac{1}{(\mathcal{D}^{(2)})^2}\right](r_2^2 + v_2).$$

Writing the above in matrix notation we obtain for $\mathbf{v} = (v_1, v_2)'$ and $\mathbf{r}^2 = (r_1^2, r_2^2)'$,

$$\mathbf{v} = c^2 M(\mathbf{v} + \mathbf{r}^2) + \mathbf{h},$$

where (note that $\mathcal{D}^{(1)} \overset{d}{=} \mathcal{D}^{(2)}$),

$$M = \frac{\mathbb{E}\left[1/(\mathcal{D}^{(1)})^2\right]}{2}\begin{bmatrix} a & b \\ b & a \end{bmatrix} \quad \text{and} \quad \mathbf{h} = \begin{bmatrix} (1-c)^2 s(1-s) \\ 0 \end{bmatrix}.$$

Moreover, use the observation that

$$M = BAB, \quad A = \frac{\mathbb{E}[1/(\mathcal{D}^{(1)})^2]}{2} \begin{bmatrix} (a-b) & 0 \\ 0 & (a+b) \end{bmatrix}, \quad B = \frac{1}{\sqrt{2}} \begin{bmatrix} -1 & 1 \\ 1 & 1 \end{bmatrix},$$

so the maximum eigenvalue of M is $\mathbb{E}[1/(\mathcal{D}^{(1)})^2]\mathbb{E}[\mathcal{D}^{(1)} - 1]$. Since for a Poisson random variable N with mean μ we have that

$$\mathbb{E}[1/(N+1)^2]E[N] = \sum_{n=1}^{\infty} \frac{e^{-\mu}\mu^n}{n \cdot n!} = \mathbb{E}\left[\frac{1}{N \vee 1}\right] - e^{-\mu} \leq 1, \qquad (13)$$

then the matrix $I - c^2 M$ is invertible, and we obtain

$$\mathbf{v} = (I - c^2 M)^{-1}(c^2 M \mathbf{r^2} + \mathbf{b}) = \sum_{k=0}^{\infty} c^{2k} M^k (c^2 M \mathbf{r^2} + \mathbf{b})$$

$$= B \begin{bmatrix} \frac{c^2 A_{11}}{1-c^2 A_{11}} & 0 \\ 0 & \frac{c^2 A_{22}}{1-c^2 A_{22}} \end{bmatrix} B\mathbf{r^2} + B \begin{bmatrix} \frac{1}{1-c^2 A_{11}} & 0 \\ 0 & \frac{1}{1-c^2 A_{22}} \end{bmatrix} B\mathbf{b}.$$

Setting $g_i = c^2 A_{ii}$ for $i = 1, 2$, and computing the product of matrices gives:

$$\mathbf{v} = \frac{1}{2(1 - g_1)(1 - g_2)} \left(K\mathbf{r^2} + (1 - c)^2 s(1 - s)\mathbf{k}\right),$$

for K and $\mathbf{k}$ defined in the statement of the lemma.

Further, if we let $\Delta_2 = 2(1 - g_1)(1 - g_2)$ and expand the above equation, we obtain

$$\mathbf{v} = \frac{1}{\Delta_2}\left(\begin{bmatrix} (g_1 + g_2 - 2g_1 g_2)r_1^2 + (-g_1 + g_2)r_2^2 \\ (-g_1 + g_2)r_1^2 + (g_1 + g_2 - 2g_1 g_2)r_2^2 \end{bmatrix} + (1-c)^2 s(1-s)\begin{bmatrix} (2 - (g_1 + g_2)) \\ (-g_1 + g_2) \end{bmatrix}\right).$$

From Eqs. (8) and (9) we also get that $r_i \leq s$ for $i = 1, 2$, so we can reduce this to

$$\mathbf{v} \leq \frac{1}{\Delta_2}\begin{bmatrix} 2g_2(1 - g_1)s^2 + (2 - (g_1 + g_2))(1 - c)^2 s(1 - s) \\ 2g_2(1 - g_1)s^2 + (-g_1 + g_2)(1 - c)^2 s(1 - s) \end{bmatrix}.$$

Plugging in $\Delta_2 = 2(1 - g_1)(1 - g_2)$, and noting that $g_2 \geq g_1$, we get

$$v_1 \leq \frac{g_2 s^2}{1 - g_2} + \frac{1}{2}\left(\frac{1}{1 - g_2} + \frac{1}{1 - g_1}\right)(1 - c)^2 s(1 - s)$$

$$\leq \frac{g_2 s^2}{1 - g_2} + \frac{1}{1 - g_2}(1 - c)^2 s(1 - s),$$

and

$$v_2 \leq \frac{g_2 s^2}{1 - g_2} + \frac{1}{2}\left(\frac{1}{1 - g_2} - \frac{1}{1 - g_1}\right)(1 - c)^2 s(1 - s)$$

$$\leq \frac{g_2 s^2}{1 - g_2} + \frac{g_2}{2(1 - g_1)(1 - g_2)}(1 - c)^2 s(1 - s).$$

Finally, using $\mathbb{E}[1/(\mathcal{D}^{(1)})^2] \leq 8/(a+b)^2$ and (13), we have $g_2 \leq \min\{c^2, 4c^2/(a+b)\}$, and so

$$v_1 \leq \frac{4c^2 s^2}{(a+b)(1-c^2)} + \frac{1-c}{1+c}s(1-s),$$

$$v_2 \leq \frac{4c^2 s^2}{(a+b)(1-c^2)}\left(1 + \frac{(1-c)(1-s)}{2s(1+c)}\right).$$

We are now ready to prove Theorem 2.

Proof (Proof of Theorem 2). For any $z > 0$, Chebyshev's inequality gives

$$\mathbb{P}(\mathcal{R}^{(1)} \leq r_1 - z) = \mathbb{P}(\mathcal{R}^{(1)} - r_1 \leq -z)$$

$$\leq \frac{v_1}{z^2}$$

$$\leq \frac{1}{z^2}\left(\frac{4c^2 s^2}{(a+b)(1-c^2)} + \frac{1-c}{1+c}s(1-s)\right). \qquad (14)$$

A similar application of Chebyshev's inequality for any $w > 0$ with $\mathcal{R}^{(2)}$ gives

$$\mathbb{P}\left(\mathcal{R}^{(2)} > \frac{s}{2} + w\right) \leq \mathbb{P}(\mathcal{R}^{(2)} > r_2 + w) \leq \frac{v_2}{w^2}$$

$$\leq \frac{1}{w^2}\frac{4c^2 s^2}{(a+b)(1-c^2)}\left(1 + \frac{(1-c)(1-s)}{2s(1+c)}\right), \qquad (15)$$

where the first inequality follows from Eq. (11). Choosing $c \in (1/2, 1-8b/(a+b)]$ results in $r_1 \geq 3s/4$, so choosing $z = w = s/8$ and plugging into the bounds from Eqs. (14) and (15) gives

$$\mathbb{P}\left(\mathcal{R}^{(1)} < \frac{5s}{8}\right) \leq \frac{256c^2}{(a+b)(1-c^2)} + \frac{64(1-c)}{1+c}\cdot\frac{1-s}{s},$$

$$\mathbb{P}\left(\mathcal{R}^{(2)} > \frac{5s}{8}\right) \leq \frac{256c^2}{(a+b)(1-c^2)}\left(1 + \frac{(1-c)(1-s)}{2s(1+c)}\right).$$

4   Results from Simulations

We illustrate the algorithm with some simulation experiments. First, we calculated the personalized PageRank scores for a 2-community dSBM with $n = 20000$ vertices, $a = 150$, $b = 10$, $s = .2$ and $c = .85$. The plot shows a clear separation of the PPR scores of the seeds, the rest of community 1 and the vertices in community 2 (Fig. 1).

We also investigated the role of the damping factor c and the best way to choose it. One natural way of doing so is to find the value of c that maximizes the difference between the mean PPR scores for the two communities. Note that

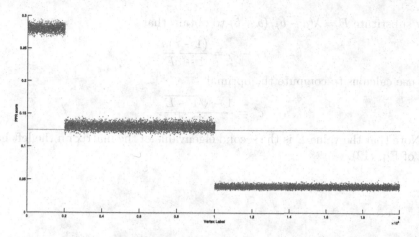

Fig. 1. A plot of the personalized PageRank scores for a 2-community dSBM with $a = 150$, $b = 10$, $n = 20000$, $s = 0.2$, and $c = .85$. The first 2000 vertices are the seeds, vertices 2001–10000 are the rest of community 1, and vertices 10001–20000 are community 2. The horizontal black line is our cutoff level $5s/8$. Proportion of misclassified community 1 vertices is 0.0935.

$r_1 - r_2$ is strictly monotone in c, but if we let $\hat{r}_1$ to be the mean of the non-seeded members of community 1, we see in Fig. 2 that $\hat{r}_1 - r_2$ is strictly convex with a maximum attained at $c = .86$. We have a description for the optimal c^* as follows.

Lemma 3. *Let $\hat{r}_1$ and r_2 be as described above. Then*

$$c^* := \operatorname{argmax}_c\{\hat{r}_1(c) - r_2(c)\} = \frac{1 - \sqrt{1 - E}}{E},$$

where

$$E = \frac{a - b}{a + b}\left(1 - e^{-(a+b)/2}\right).$$

Proof. To calculate $\hat{r}_1$, we consider the dSBM to have 3 communities, where we separate the seeds and the rest of the vertices in community 1. Then, Theorem 1 gives us a system of 3 distributional fixed-point equations. Using those, and calculations similar to the ones we did for Lemma 1, we get

$$\hat{r}_1 = (1 - c)s\left(\frac{1 - \frac{c\lambda a}{a+b}}{(1 - c\lambda)\left(1 - c\lambda\left(\frac{a-b}{a+b}\right)\right)} - 1\right) \tag{16}$$

$$r_2 = (1 - c)s\left(\frac{\frac{c\lambda b}{a+b}}{(1 - c\lambda)\left(1 - c\lambda\frac{a-b}{a+b}\right)}\right).$$

162   S. Banerjee et al.

Now substitute $E = \lambda(a - b)/(a + b)$ to obtain that

$$\hat{r}_1 - r_2 = \frac{(1 - c)s}{1 - cE},$$

and use calculus to compute the optimal

$$c^* = \frac{1 - \sqrt{1 - E}}{E}.$$

Note that the value E is the second eigenvalue of the matrix on the left hand side of Eq. (12).

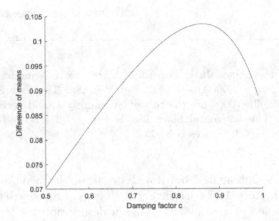

Fig. 2. Plot of $\hat{r}_1 - r_2$ as c varies from .5 to 1 for a smaller dSBM with $n = 2000$, $a = 100$, $b = 2$ and $s = .15$.

5   Remarks and Conclusions

In the sparse regime, we have proposed a cutoff level to identify vertices of community 1 based on their personalized PageRank scores and provided theoretical bounds on the probability of misclassifying a vertex. Our bounds are not tight, and simulations indicate that we might be able to use a lower threshold to further reduce the error (see also Remark 2). Another possible threshold option in the case of the symmetric SBM ($\pi_1 = \pi_2$) is the median of PageRank scores. We also believe that the proposed method should work for asymmetric dSBMs with $\pi_1 \neq \pi_2$, but the expressions for the mean and variance of PageRank become too complicated to compute clean bounds. Possible future work could include trying to show that the π_1-th quantile of the limiting PageRank distribution is a good threshold in the case $\pi_1 \neq \pi_2$, or trying to find a threshold independent of π so that we can recover communities even when we do not have information about their sizes. Another interesting direction would be to investigate whether the inference can be strengthened if the seed set contains members from both communities and/or the connectivity structure of the subgraph spanned by the seeds is fully or partially known.

References

1. Avrachenkov, K., Kadavankandy, A., Litvak, N.: Mean field analysis of personalized PageRank with implications for local graph clustering. J. Stat. Phys. **173**(3), 895–916 (2018)
2. Garavaglia, A., van der Hofstad, R., Litvak, N.: Local weak convergence for PageRank. Ann. Appl. Probab. **30**(1), 40–79 (2020)
3. Gulikers, L., Lelarge, M., Massoulié, L.: An impossibility result for reconstruction in the degree-corrected stochastic block model. Ann. Appl. Probab. **28**(5), 3002–3027 (2018)
4. Kloumann, I.M., Ugander, J., Kleinberg, J.: Block models and personalized PageRank. Proc. Nat. Acad. Sci. **114**(1), 33–38 (2017)
5. Lee, J., Olvera-Cravioto, M.: PageRank on inhomogeneous random digraphs. Stoch. Process. Appl. **130**(4), 2312–2348 (2020)
6. Mossel, E., Neeman, J., Sly, A.: Consistency thresholds for binary symmetric block models. arXiv preprint arXiv:1407.1591, vol. 3, no. 5 (2014)
7. Mossel, E., Neeman, J., Sly, A.: Reconstruction and estimation in the planted partition model. Probab. Theor. Relat. Fields **162**(3), 431–461 (2015)
8. Massoulié, L.: Community detection thresholds and the weak Ramanujan property. In: Proceedings of the Forty-Sixth Annual ACM Symposium on Theory of Computing, pp. 694–703 (2014)
9. Chin, P., Rao, A., Vu, V.: Stochastic block model and community detection in sparse graphs: a spectral algorithm with optimal rate of recovery. In: Conference on Learning Theory, pp. 391–423. PMLR (2015)
10. Spielman, D.A., Teng, S.H.: A local clustering algorithm for massive graphs and its application to nearly linear time graph partitioning. SIAM J. Comput. **42**(1), 1–26 (2013)
11. Andersen, R., Chung, F., Lang, K.: Local partitioning for directed graphs using PageRank. In: Bonato, A., Chung, F.R.K. (eds.) WAW 2007. LNCS, vol. 4863, pp. 166–178. Springer, Heidelberg (2007). https://doi.org/10.1007/978-3-540-77004-6_13
12. Dhara, S., Gaudio, J., Mossel, E., Sandon, C.: Spectral recovery of binary censored block models. In: Proceedings of the 2022 Annual ACM-SIAM Symposium on Discrete Algorithms (SODA), pp. 3389–3416. Society for Industrial and Applied Mathematics (2022)
13. Das Sarma, A., Molla, A.R., Pandurangan, G., Upfal, E.: Fast distributed PageRank computation. In: Frey, D., Raynal, M., Sarkar, S., Shyamasundar, R.K., Sinha, P. (eds.) ICDCN 2013. LNCS, vol. 7730, pp. 11–26. Springer, Heidelberg (2013). https://doi.org/10.1007/978-3-642-35668-1_2
14. Olvera-Cravioto, M.: PageRank's behavior under degree correlations. Ann. Appl. Probab. **31**(3), 1403–1442 (2021)

A Simple Model of Influence

Colin Cooper[✉], Nan Kang[✉], and Tomasz Radzik[✉]

Department of Informatics, King's College London, London, UK
tomasz.radzik@kcl.ac.uk

Abstract. We propose a simple model of influence in a network, based on edge density. In the model vertices (people) follow the opinion of the group they belong to. The opinion percolates down from an active vertex, the influencer, at the head of the group. Groups can merge, based on interactions between influencers (i.e., interactions along 'active edges' of the network), so that the number of opinions is reduced. Eventually no active edges remain, and the groups and their opinions become static.

Our analysis is for $G(n, m)$ as m increases from zero to $N = \binom{n}{2}$. Initially every vertex is active, and finally G is a clique, and with only one active vertex. For $m \leqslant N/\omega$, where $\omega = \omega(n)$ grows to infinity, but arbitrarily slowly, we prove that the number of active vertices $a(m)$ is concentrated and we give w.h.p. results for this quantity. For larger values of m our results give an upper bound on $\mathbb{E}\, a(m)$.

We make an equivalent analysis for the same network when there are two types of influencers. Independent ones as described above, and stubborn vertices (dictators) who accept followers, but never follow. This leads to a reduction in the number of independent influencers as the network density increases. In the deterministic approximation (obtained by solving the deterministic recurrence corresponding to the formula for the expected change in one step), when $m = cN$, a single stubborn vertex reduces the number of influencers by a factor of $\sqrt{1-c}$, i.e., from $a(m)$ to $(\sqrt{1-c})\, a(m)$. If the number of stubborn vertices tends to infinity slowly with n, then no independent influencers remain, even if $m = N/\omega$.

Finally we analyse the size of the largest influence group which is of order $(n/k) \log k$ when there are k active vertices, and remark that in the limit the size distribution of groups is equivalent to a continuous stick breaking process.

Keywords: Random graphs and processes · Social networks and influence

1 Introduction

We propose a simple model of influence in a network, based on edge density. In the model vertices (people) follow the opinion of the group they belong to. This opinion percolates down from an active (or opinionated) vertex, the influencer, at the head of the group. Groups can merge, based on edges between influencers (active edges), so that the number of opinions is reduced. Eventually no active edges remain and the groups and their opinions become static.

© The Author(s), under exclusive license to Springer Nature Switzerland AG 2023
M. Dewar et al. (Eds.): WAW 2023, LNCS 13894, pp. 164–178, 2023.
https://doi.org/10.1007/978-3-031-32296-9_11

The sociologist Robert Axelrod [1] posed the question "If people tend to become more alike in their beliefs, attitudes, and behavior when they interact, why do not all such differences eventually disappear?". This question was further studied by, for example, Flache et al. [4], Moussaid et al. [7] who review various models of social interaction, generally based on some form of agency.

In our model the emergence of separate groups occurs naturally due to lack of active edges between influencers. The exact composition of the groups and their influencing opinion being a stochastic outcome of the connectivity of individual vertices.

Joining Protocol. The process models how networks can partition into disjoint subgraphs which we call *fragments* based on following the opinion of a neighbour. At any step, a fragment consists of a directed tree rooted at an active vertex (the influencer), edges pointing from follower vertices towards the root. This forms a simple model of influence where the vertices in a fragment follow the opinion of the vertex they point to, and hence that of the active root.

The process is carried out on a fixed underlying graph $G = (V, E)$. The basic u.a.r. process is as follows.

1. Vertices are either active or passive. Initially all vertices of V are active, and all fragments are individual vertices.
2. (a) *Vertex model*: An active vertex u is chosen u.a.r. and contacts a random active neighbour v.
   (b) *Edge model*: A directed edge (v, u) between active vertices is chosen u.a.r. and the active vertex u contacts its active neighbour v. (Equivalently, an undirected edge is chosen u.a.r. and random of the two vertices contacts the other vertex).
3. The contacted neighbour v becomes passive.
4. Vertex v directs an edge to u in the fragment graph. Vertex v and its fragment $F'(v)$ become part of the fragment $F(u)$ rooted at u.
5. An active vertex is isolated if it has no edges to active neighbours in G. The process ends when all active vertices are isolated.

Summary of Results. As an illustration of the process, in this paper we make an analysis of the edge model for random graphs $G(n, m)$, providing the following results.

- Theorem 1 gives the w.h.p. number of fragments in $G(n, m)$ for $m \ll \binom{n}{2}$, and an upper bound on the expected number for any m. The results are supported by simulations which indicate that the upper bound in Theorem 1 is the correct answer.
- Theorem 2 gives the equivalent number of fragments in the presence of stubborn vertices (vertices who accept followers, but refuse to follow).
- The tail distribution of size of the largest fragment and its expected size are given in Lemma 3.

2   Analysis for Random Graphs $G(n, m)$

We suppose the underlying graph is a random graph $G(n, m)$ and at each step, the absorbing vertex and the contacted neighbour are chosen by selecting uniformly at random an edge between two active vertices (the edge model).

We work with a random permutation σ of the edges of the complete graph, where $N = \binom{n}{2}$. The edges of σ are inspected in the permutation order. By revealing the first m edges in the random permutation we choose a random graph $G(n, m)$. The order in which we reveal these first m edges and their random directions give a random execution of the joining protocol on the chosen graph.

The u.a.r. process is equivalent to picking a random edge between active vertices (by skipping the steps when one or both vertices of the chosen edge is not active). One endpoint stays active and the other becomes passive. It doesn't matter which (since we are interested in the number and sizes of fragments, but not in their structure). As none of the edges between active vertices have been inspected, the next edge is equally likely to be between any of them.

Let $A(m)$ be the set of active vertices obtained by running the process on $G(n, m)$. Let $a(m) = |A(m)|$ be the number of active vertices after m edges are exposed.

The following deterministic recurrence plays a central part in our analysis,

$$a_{t+1} = a_t - \frac{a_t(a_t - 1)}{2(N - t)}. \tag{1}$$

We will show that if $t = o(N)$ then $\mathbb{E}\, a(t) \sim a_t$, and that $\mathbb{E}\, a(t) \leqslant a_t$ always. The solution to (1) is given in the next lemma. To maintain continuity of presentation, the proof of the lemma is deferred to the next section.

In what follows ω is a generic variable which tends to infinity with n, but can do so arbitrarily slowly.

Lemma 1.

(i) For $t \leqslant N/\omega$, we have $a_t = c_t(1 + \varepsilon_t)$ for $\varepsilon_t = O(t/(N - t))$ and c_t given by

$$c_t = \frac{n^2}{n + t}. \tag{2}$$

Thus if $t = o(N)$, $a_t \sim c_t$.

(ii) For $t \leqslant N - \omega$, we have $a_t = b_t(1 + \varepsilon_t)$ for $\varepsilon_t = O(1/\omega)$ and b_t given by

$$b_t = \frac{1}{1 - (1 - 1/n)\sqrt{1 - t/N}}. \tag{3}$$

Our first result follows from this lemma.

Theorem 1. *Given a random graph $G(n, m)$, for $m \leqslant N/\omega$, where $\omega \to \infty$, the number of components $a(m)$ generated by the opinion fragmentation process is concentrated with expected value given by*

$$\mathbb{E}\, a(m) \sim \frac{1}{1 - (1 - 1/n)\sqrt{1 - m/N}}. \tag{4}$$

Moreover for any $m \leqslant N(1 - o(1))$, $\mathbb{E}\, a(m)$ is upper bounded by the RHS of (4).

For $m \leqslant N/\omega$, the expected number of active vertices $\mathbb{E}\, a(m)$ is well approximated by the simpler expression $\mathbb{E}\, a(m) \sim \frac{n^2}{n+m}$ as given by (2).

Proof. We add edges of a complete graph to an empty graph in random order, and analyse the expected change in the number of active vertices in one step. At the beginning all vertices are active and $a(0) = n$.

Let $a(t)$ be the total number of active vertices at step t. There are $N - t$ unexamined edges remaining after step t as we add one edge per step, and there are $\binom{a(t)}{2}$ many active edges left after step t. Therefore, the probability of choosing an active edge at step $t + 1$ is $\binom{a(t)}{2}/(N - t)$, and we lose one active vertex for each active edge added. Thus,

$$\mathbb{E}\,[a(t+1) \mid a(t)] = a(t) - \frac{\binom{a(t)}{2}}{N - t}. \tag{5}$$

The function x^2 is convex so $\mathbb{E}\,(a(t))^2 \geqslant (\mathbb{E}\, a(t))^2$. Thus the solution a_t of the recurrence (1) gives an upper bound on $\mathbb{E}\, a(t)$.

On the other hand, if $a(t)$ is concentrated, then $\mathbb{E}\,(a(t))^2 \sim (\mathbb{E}\, a(t))^2$ in which case $\mathbb{E}\, a(t) \sim a_t$ as in (1). This is easy up to $t \leqslant n^{4/3}/\omega$. Using a edge exposure martingale, the value of $a(t)$ can only change by zero or one at any step, so

$$\mathbb{P}(|a(t) - \mathbb{E}\, a(t)| \geqslant \lambda) \leqslant \exp\left(-\frac{\lambda^2}{2t}\right). \tag{6}$$

For $t \leqslant n^{4/3}/\omega$, choose $\lambda = \sqrt{\omega t}$ to get $o(1)$ on the RHS in (6) and $\lambda = n^{2/3} = o(a_t)$. Assuming concentration of $a(t)$, $\mathbb{E}\,(a(t))^2$ on the RHS of (5) can be replaced by $(1 + o(1))(\mathbb{E}\, a(t))^2$. This allows us to use recurrence (1) to analyse the recurrence (5) for $\mathbb{E}\, a(t)$.

From $t \geqslant n^{4/3}$ onward, mostly nothing happens at any step and the standard Azuma-Hoeffding inequality approach stops working. As $a(t)$ is a supermartingale ($\mathbb{E}\, a(t+1) \leqslant a(t)$), we can use Freedman's inequality, which we paraphrase from [2].

FREEDMAN'S INEQUALITY [5]. *Suppose $Y_0, Y_1, \ldots$ is a supermartingale such that $Y_j - Y_{j-1} \leqslant C$ for a positive constant C and all j. Let $V_m = \sum_{k \leqslant m} \mathrm{Var}[(Y_{k+1} - Y_k) \mid \mathcal{F}_k]$. Then for $\lambda, b > 0$*

$$\mathbb{P}(\exists m : V_m \leqslant b \text{ and } Y_m - \mathbb{E}\, Y_m \geqslant \lambda) \leqslant \exp\left(-\frac{\lambda^2}{2(b + C\lambda)}\right). \tag{7}$$

In our case $Y_m = a(m + s)$ given the value of $a(s)$, and $C = 1$ by (5).

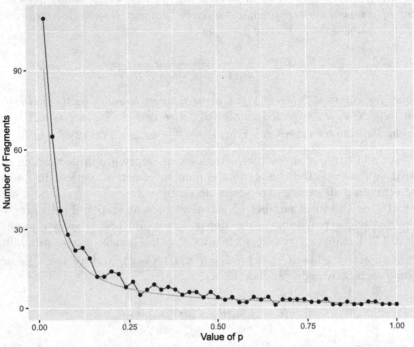

Fig. 1. Simulation of the number of active vertices. The simulation is based on $G(n,p)$. For large $m = Np$, $G(n,p) \sim G(n,m)$ so the results are equivalent. The blue plot is the simulation, with values of p interpolated at 0.02, the first entry being $p = 0.02$. The red curve is (3) giving b_m evaluated at $p = m/N$. (Color figure online)

Let $t_1 = n^{4/3}/\omega = n^{1+1/3}/\omega$ and $t_i = n^{1+1/3+\cdots+2^{i-1}/3^i}/\omega = n^{1+\beta_i}/\omega$, where $\beta_i = 1 - 2^i/3^i$ and ω may vary. The inductive assumption is that

$$a(t_i) \sim \frac{n^2}{n+t_i} \sim \omega n^{1-\beta_i},$$

see Lemma 1. As $a(t)$ is monotone non-increasing it follows for $t_i \leqslant t \leqslant t_{i+1}$, and $t_i = o(N)$, that

$$\frac{\binom{a(t)}{2}}{N-t} \leqslant \frac{a(t_i)^2}{n^2}(1+o(1)) \sim \frac{\omega^2 n^{2-2\beta_i}}{n^2} = \frac{\omega^2}{n^{2\beta_i}}.$$

As $\mathbb{V}ar(a(t+1) - a(t)) \leqslant (1+o(1))a(t_i)^2/n^2$ we have that

$$b_i = \sum_{t=t_i}^{t_{i+1}} \mathbb{V}ar(a(t+1) - a(t)) \leqslant (1+o(1))\frac{t_{i+1}}{n^{2\beta_i}} = \omega n^{1-\beta_i+2^i/3^{i+1}}.$$

Thus using (7),

$$\mathbb{P}(a(t_{i+1}) - \mathbb{E}\,a(t_{i+1}) \geqslant \omega^{3/4} n^{1-\beta_{i+1}})$$

$$\leqslant \exp\left(-\frac{1}{2}\omega^{6/4}\frac{n^{2-2\beta_{i+1}}}{\omega n^{1-\beta_i+2^i/3^{i+1}}(1+o(1))}\right)$$

$$= \exp\left(-\frac{1}{2}\omega^{1/2}n^{1-\beta_{i+1}-2^{i+1}/3^{i+1}}\right) = \exp\left(-\frac{1}{2}\omega^{1/2}\right).$$

The last line follows because $\beta_{i+1} = 1-2^{i+1}/3^{i+1}$. For simplicity let $\omega = C\log^2 n$ for some large constant C. As $i \to \infty$, β_i tends to one, and we have that w.h.p. $a(t) \sim \mathbb{E}\,a(t)$ for any $t = o(N) = N/\omega$ say. Hence $\mathbb{E}\,(a(t)^2) \leqslant (1+o(1))(\mathbb{E}\,a(t))^2$. From an earlier part of this theorem, $\mathbb{E}\,(a(t))^2 \geqslant (\mathbb{E}\,a(t))^2$. This completes the proof of the theorem.

Simulation results (see Fig. 1), suggest that $\mathbb{E}\,a(t)$ should continue to track b_t of (3) throughout.

3   Proof of Lemma 1

To solve (5), the first step is to solve the equivalent deterministic recurrence (1), i.e.,

$$a_{t+1} = a_t\left(1 - \frac{a_t - 1}{2(N-t)}\right). \tag{8}$$

An approximate solution can be obtained by replacing a_t in this recurrence by a differential equation in $b(t)$. The initial condition $b(0) = a_0 = n$ gives

$$\frac{d\,b(t)}{dt} = -\frac{\binom{b(t)}{2}}{N-t} \quad\Longrightarrow\quad b(t) = \frac{1}{1 - \left(1-\frac{1}{n}\right)\sqrt{1-\frac{t}{N}}}. \tag{9}$$

We now prove Lemma 1, restating it below for convenience.

Lemma 2.

(i) Let c_t be given by,

$$c_t = \frac{n^2}{n+t}.$$

For $t = N/\omega$, we have $a_t = c_t(1+\varepsilon_t)$ for $\varepsilon_t = O(t/(N-t))$. Thus if $t = o(N)$, $a_t \sim c_t$.

(ii) For $t \leqslant N - \omega$, where $\omega \to \infty$, we have $a_t = b_t(1+\varepsilon_t)$ for $\varepsilon_t = O(1/\omega)$ and b_t given by

$$b_t = \frac{1}{1 - (1-1/n)\sqrt{1-t/N}}.$$

Proof. (i) We define ε_t so that $a_t = c_t(1 + \varepsilon_t)$, and show by induction that for $0 \leqslant t \leqslant N/\omega$, we have $-2t/(N-t) \leqslant \varepsilon_t \leqslant 0$, starting from $a_0 = c_0 = n$ and $\varepsilon_0 = 0$. We take an arbitrary $0 \leqslant t \leqslant N/\omega$, and evaluate the recurrence (8) for a_{t+1}, assuming inductively that $a_t = c_t(1 + \gamma t/(N-t))$ for some $\gamma \in [-2, 0]$.

$$
\frac{a_t - 1}{2(N-t)} = \frac{\frac{n^2}{n+t}\left(1 + \frac{\gamma t}{N-t}\right) - 1}{2(N-t)} = \frac{n^2 - n - t}{2(n+t)(N-t)} + \frac{\gamma t n^2}{2(n+t)(N-t)^2}
$$

$$
= \frac{2(N-t) + t}{2(n+t)(N-t)} + \frac{\gamma t n^2}{2(n+t)(N-t)^2}
$$

$$
= \frac{1}{n+t} + \frac{t}{2(n+t)(N-t)} + \frac{\gamma t n^2}{2(n+t)(N-t)^2}.
$$

Thus

$$
\frac{a_{t+1}}{c_{t+1}} = \frac{c_t}{c_{t+1}}\left(1 + \frac{\gamma t}{N-t}\right)\left(1 - \frac{1}{n+t} - \frac{t}{2(n+t)(N-t)} - \frac{\gamma t n^2}{2(n+t)(N-t)^2}\right)
$$

$$
= \left(1 + \frac{1}{n+t}\right)\left(1 + \frac{\gamma t}{N-t}\right)
$$

$$
\times \left(1 - \frac{1}{n+t} - \frac{t}{2(n+t)(N-t)} - \frac{\gamma t n^2}{2(n+t)(N-t)^2}\right)
$$

$$
= 1 + \frac{\gamma t}{N-t} - \frac{t}{2(n+t)(N-t)} - \frac{\gamma t n^2}{2(n+t)(N-t)^2} + \delta = 1 + \xi + \delta,
$$

where $-1/(N-t) \leqslant \delta \leqslant 1/(2(N-t))$, by inspecting the terms contributing to δ and using the assumption that $t \leqslant N/\omega$. Now we have, recalling that $-2 \leqslant \gamma \leqslant 0$,

$$
\xi = \frac{\gamma t}{N-t}\left(1 - \frac{n^2}{2(n+t)(N-t)}\right) - \frac{t}{2(n+t)(N-t)} \leqslant -\frac{1}{2(N-t)}
$$

and

$$
\xi \geqslant -\frac{2t}{N-t} - \frac{t}{2(n+t)(N-t)}
$$

$$
\geqslant -\frac{2(t+1)}{N-t-1} + \frac{2N}{(N-t-1)(N-t)} - \frac{t}{2(n+t)(N-t)}
$$

$$
\geqslant -\frac{2(t+1)}{N-t-1} + \frac{1}{N-t}\left(\frac{2N}{N-t-1} - \frac{t}{2(n+t)}\right) \geqslant -\frac{2(t+1)}{N-t-1} + \frac{1}{N-t}.
$$

The bounds on ξ and δ imply that $a_{t+1}/c_{t+1} = 1 + \gamma'(t+1)/(N-t-1)$, for some $-2 \leqslant \gamma' \leqslant 0$.

(ii) We note firstly that $c_t \geqslant b_t$ and establish that for $t = o(N)$, $c_t \sim b_t$. Let $t_1 \leqslant N/\omega$. Using $\sqrt{1-x} = 1 - x/2 - x^2/4 - O(x^3)$, it can be checked that

$$
\frac{1}{1 - (1 - 1/n)\sqrt{1 - t_1/N}} = (1 + \delta)\frac{n^2}{n + t_1},
$$

where $\delta = O(t_1^2/n^3(n+t_1)) = O(1/\omega^2)$. Thus $a_{t_1} = b_{t_1}(1 + O(1/\omega))$.

Let $\theta = (n-1)/n$. Assume $N - t \geqslant \omega \to \infty$. Then

$$\frac{1}{b_{t+1}} = 1 - \theta\sqrt{1 - \frac{t+1}{N}} = 1 - \theta\sqrt{1 - \frac{t}{N}}\sqrt{1 - \frac{1}{N-t}}$$

$$= 1 - \theta\sqrt{1 - \frac{t}{N}}\left(1 - \frac{1}{2(N-t)} - \frac{1}{4(N-t)^2}(1 + O(1/\omega))\right),$$

and so

$$\frac{b_t}{b_{t+1}} = 1 + \frac{\theta\sqrt{1 - \frac{t}{N}}}{2(N-t)(1 - \theta\sqrt{1 - \frac{t}{N}})} + \frac{\theta\sqrt{1 - \frac{t}{N}}}{4(N-t)^2(1 - \theta\sqrt{1 - \frac{t}{N}})}(1 + o(1)).$$

Let

$$\lambda = \frac{\theta\sqrt{1 - \frac{t}{N}}}{2(N-t)(1 - \theta\sqrt{1 - \frac{t}{N}})},$$

then also $\lambda = (b_t - 1)/(2(N-t))$. Thus

$$a_{t+1} = a_t\left(1 - \frac{a_t - 1}{2(N-t)}\right)$$

$$= b_t(1 + \varepsilon_t)\left(1 - \frac{(b_t - 1)(1 + \varepsilon_t) + \varepsilon_t}{2(N-t)}\right)$$

$$= b_{t+1}\left(1 + \lambda + \frac{\lambda}{2(N-t)}(1 + o(1))\right)(1 + \varepsilon_t)\left(1 - \lambda(1 + \varepsilon_t) - \frac{\varepsilon_t}{2(N-t)}\right)$$

$$= b_{t+1}\left(1 + \varepsilon_t - \lambda^2 - \frac{\lambda}{2(N-t)} + O\left(\varepsilon_t\left(\lambda + \frac{1}{N-t}\right)\right)\right).$$

It follows from $\theta\sqrt{1-x} < \sqrt{1-x} \leqslant 1 - x/2$, that $b_x \leqslant 2/x$. Thus $b_t \leqslant 2N/t$, $\lambda \leqslant N/(t(N-t))$ and

$$\lambda^2 \leqslant \frac{N}{t^2(N-t)}, \qquad \frac{\lambda}{N-t} \leqslant \frac{N}{t(N-t)^2}.$$

Finally with $t_1 = N/\omega$ as above

$$|\varepsilon_t| \leqslant |\varepsilon_{t_1}| + \sum_{t_1}^{t}\lambda^2 + \frac{\lambda}{N-t} \sim \int_{t_1}^{t}\frac{N}{t^2(N-t)} + \frac{N}{t(N-t)^2} = \int F(t)dt.$$

Denote $t_1 = c_1 N$ and $t_2 = c_2 N$ where $t_2 \leqslant N - \omega$. Then

$$\int F(t)dt = \int \frac{1}{t^2} + \frac{2}{Nt} - \frac{2}{N(N-t)} + \frac{3}{(N-t)^2}$$

$$= \left[-\frac{1}{t} + \frac{2}{N}\log t + \frac{2}{N}\log(N-t) + \frac{3}{N-t}\right]_{Nc_1}^{Nc_2}$$

$$\leqslant \frac{1}{N}\left(\frac{1}{c_1} + 2\log\frac{c_2}{c_1} + 2\log\frac{1-c_2}{1-c_1} + \frac{1}{1-c_2}\right),$$

and thus if $c_2 = 1 - \omega/N$, from the last term,

$$|\varepsilon_{t_2}| \leqslant O\left(\frac{1}{\omega}\right).$$

4   The Effect of Stubborn Vertices

A vertex is stubborn (intransigent, autocratic, dictatorial) if it holds fixed views, and although happy to accept followers, it refuses to follow the views of others. Typical examples include news networks, politicians and some cultural or religious groups. Stubborn vertices can only be root vertices.

We note that voting in distributed systems in the presence of stubborn agents has been extensively studied see e.g., [8,10,11] and references therein.

The effect of stubborn vertices on the number of other active vertices in the network depends on the edge density, as is illustrated by the next theorem. Let $a_k(t)$ be the number of *active independent vertices* at step t in the presence of $k \geqslant 0$ stubborn vertices. As the stubborn vertices are never absorbed, the total number of roots is $a_k(t) + k$.

Let $\beta = 2k - 1$, and

$$b_k(t) \sim \beta \, \frac{\left(\frac{n}{n+\beta}\right)\left(1 - \frac{t}{N}\right)^{\beta/2}}{1 - \left(\frac{n}{n+\beta}\right)\left(1 - \frac{t}{N}\right)^{\beta/2}}. \tag{10}$$

Essentially we solve the deterministic recurrence equivalent to (1) to obtain (10), and argue by concentration, convexity and super-martingale properties that $b_1(t)$ is the asymptotic solution ($t = o(N)$) or an effective upper bound ($t \leqslant N$). Due to space limitations the proof is only given in outline.

Theorem 2. *(i)* **One stubborn vertex.** *Let* $N = \binom{n}{2} + n$, *then provided* $t \leqslant N/\omega$, *the number of independent active vertices* $a_1(t) \sim a_0(t) \sim b_t$, *w.h.p., where* b_t *is the solution to* (1) *as given by* (3). *If* $t = cN$ *then* $\mathbb{E}\, a_1(cN) \leqslant b_1(cN) \leqslant (\sqrt{1-c})\, b_t$.

(ii) **A constant number** k **of stubborn vertices.** *Let* $k \geqslant 1$ *be integer, and* $N = k + \binom{n}{2}$. *If* k *is constant, and* $t \leqslant N/\omega$ *then* $a_k(t) \sim b_k(t)$ *w.h.p., and for any* $t \leqslant N$, $\mathbb{E}\, a_k(t) \leqslant b_k(t)(1 + o(1))$.

(iii) **The number** k **of stubborn vertices is unbounded.** *If* $t = N/\omega$ *and* $\omega/k \to 0$, *then w.h.p no independent active vertices are left by step* t.

Proof. To formulate the model, we note that, at the end of step t there are $ka_k(t)$ edges between stubborn and independent active vertices. Writing $a = a_k(t)$ we extend (5) with $N = \binom{n}{2} + k$ to

$$\mathbb{E}\, a_k(t + 1) = a_k(t) - \frac{ka_k}{N - t} - \frac{a_k(a_k - 1)}{2(N - t)}. \tag{11}$$

Solving the equivalent differential equation we obtain (10)

In case (iii), let $t = N/\omega$ and $k = \lambda\omega$ where $\lambda \to \infty$, but $k = o(n)$. Then

$$(1 - t/N)^{\beta/2} \sim e^{-t(2k-1)/2N} = e^{-(k-1/2)/\omega} \sim e^{-\lambda} = o(1).$$

Thus (10) tends to $b_1(t) \sim o(1)/(1 - o(1))$ and the result follows.

We remark that if the network is sparse ($c = o(1)$), and there are only a few stubborn vertices, these will have little effect. However, if the network is dense (c is a positive constant), there are fewer independent active vertices, even if k is constant. On the other hand if $k \to \infty$ even in sparse networks where $t = N/\omega$, the number of independent active vertices can tend to zero. This indicates in a simplistic way the effect of edge density (increasing connectivity) in social networks on the formation of independent opinions in the presence of vertices with fixed views. It also indicates that even in sparse networks, a large number of stubborn vertices can lead to the suppression of independent opinion formation.

Figure 2, illustrates the above Theorem. The plots show the number of active vertices in the presence of stubborn vertices (dictators). The number k of stubborn vertices is equal to 1 in the left hand plot and 5 in the right hand plot. The plots are based on $G(n, p)$, for $n = 1000$ and $p \geqslant 0.1$. The upper curve in the right hand figure is $b_k(t) + k$, the total number of active vertices. The middle curve is b_t from (3). The simulation plot marked by $+$ symbols is the final number of active vertices in a system without stubborn vertices, as in Fig. 1. The lower curve is $b_k(t)$, and the associated simulation is the number of independent active vertices in the presence of dictators.

In the left hand plot for $k = 1$, the curves $k + b_k(t)$ and b_t as given by (3) are effectively identical, so a distinct upper curve is missing. The lower curve is $b_1(t)$, and its associated plot is the number of independent active vertices in the presence of stubborn vertices.

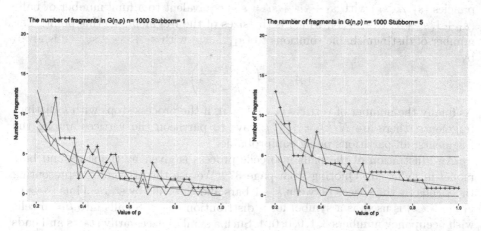

Fig. 2. The number of active vertices in the presence of stubborn vertices (dictators). The left hand plot is for $k = 1$ stubborn vertices, and the right hand plot for $k = 5$ stubborn vertices. The plots are based on $G(n, p)$, for $n = 1000$ and $p \geqslant 0.1$.

5   The Largest Fragment in $G(n, m)$

Let $F_{(1)}(m)$ denote the size of the largest fragment in $G(n, m)$. The value of $F_{(1)}(m)$, the number of followers of the dominant influencer, (we assume the influencer follows themself), will depend on the number of active vertices $a(m)$. As both $a(m)$ and $F_{(1)}$ are random variables, it is easier to fix $a(m) = k$, and study $\mathbb{E} F_{(1)} = \mathbb{E}(F_{(1)} \mid k)$ for a given value k. In the limit as $n \to \infty$, $\mathbb{E} F_{(1)}$ converges to a continuous process known as stick breaking.

The first step is to describe a consistent discrete model. This can be done in several ways, as a multivariate Polya urn, as the placement of $n - k$ unlabelled balls into k labelled boxes, or as randomly choosing $k - 1$ distinct vertices from $\{2, ..., n\}$ on the path $1, 2, ..., n$. The latter corresponds to the limiting stick breaking process.

Looking Backwards: A Polya urn Process. If we stop the process when there are exactly k active vertices for the first time, then at the previous step there were $k + 1$ active vertices. Let the k active vertices be $A_k = \{v_1, ..., v_k\}$, and let $A_{k+1} = \{v_1, ..., v_k, b\}$ be the active vertices at the previous step, where the vertex b was absorbed. As the edges $bv_1, ...bv_k$ are equiprobable, the probability b was absorbed by v_i is $1/k$.

Working backwards from k to n is equivalent to a k-coloured Polya urn, in which at any step a ball is chosen at random and replaced with 2 balls of the same colour. At the first step backwards any one of the colours $1, ..., k$ is chosen and replaced with 2 balls of the chosen colour (say colour i). This is equivalent to the event that vertex b attaches to the active vertex v_i.

Starting with k different coloured balls and working backwards for s steps is equivalent to placing s unlabelled balls into k cells. Thus any vector of occupancies $(s_1, ..., s_k)$ with $s_1 + \cdots + s_k = s$ is equivalent to a final number of balls $(s_1 + 1, s_2 + 1, ..., s_k + 1)$; which is the sizes of the fragments at this point. The number of distinguishable solutions to $(s_1, ..., s_k)$ with $s_1 + \cdots + s_k = s$ is given by

$$A_{s,k} = \binom{s + k - 1}{k - 1}.$$

As finally the number of vertices is $s + k = n$, if the process stops with k distinct fragments, there are $N(k) = \binom{n-1}{k-1}$ ways to partition the vertices among the fragments, all partitions being equiprobable.

An illustration of the balls into cells process is given by the stars and bars model in Feller [3]. Quoting from page 37, 'We use the artifice of representing the k cells by the spaces between $k + 1$ bars and the balls by stars. Thus $|***|*|||*|****|$ is used as a symbol for a distribution of $s = 8$ balls into $k = 6$ cells with occupancy numbers $3, 1, 0, 0, 0, 4$. Such a symbol necessarily starts and ends with a bar but the remaining $k - 1$ bars and s stars can appear in an arbitrary order.'

That this is equivalent to the above Polya urn model can be deduced from $A_{s+1,k}/A_{s,k} = (s + k)/(s + 1)$. The numerator is the number of positions for the extra ball. Picking a left hand bar corresponds to picking one of the k root vertices. The denominator is the number of ways to de-identify the extra ball; being the number of symbols (urn occupancies) which map to the new occupancy.

The sizes of the fragments can also be viewed as follows. Consider the path $1, 2, 3, ..., n$ with the first fragment starting at vertex 1, the left hand bar. The choice of $k-1$ remaining start positions (internal bars) from the vertices $2, 3, ..., n$ divides the path into k pieces whose lengths are the fragment sizes. Taking the limit as $n \to \infty$ and re-scaling the path length to 1, we obtain the limiting process, known as *stick breaking*.

Limiting Process: Stick Breaking. The continuous limit as $n \to \infty$ also arises as a "stick breaking" process. Let $F_n(i), i = 1, ..., k$ be the number of balls of colour i when the urn contains n balls. Then $S(i) = F_n(i)/n$ tends to the length of the i-th fragment when the unit interval is broken into k pieces using $k-1$ independent variates U uniformly distributed in $[0, 1]$. This kind of random partitioning process corresponds to a stick-breaking or spacing process in which a stick is divided into $k + 1$ fragments. The distribution of the largest fragment is well-studied [6, 9].

Lemma 3. *[6, 9] Suppose a stick of length 1 is broken into k fragments uniformly at random. Let $S_{(1)} \leqslant S_{(2)} \leqslant \ldots S_{(k)}$ be the size of these fragments given in increasing order of size. Then, for $i \in \{1, \ldots, k\}$,*

$$\mathbb{E}\, S_{(i)} = \frac{1}{k} \sum_{j=0}^{i-1} \frac{1}{k-j}.$$

Thus, the largest fragment has size $\mathbb{E}\, S_{(k)} = H_k \sim \frac{1}{k} \log k$.
The distribution of $S_{(k)}$ satisfies

$$\mathbb{P}(S_{(k)} \geqslant x) \sim \exp\left(-k e^{-kx}\right), \tag{12}$$

Thus $\mathbb{E}\, F_{(1)}$, the expected size of the largest fragment among k tends to $\mathbb{E}\, S_{(k)} = \frac{n}{k} \log k$ (Fig. 3).

Maximum Fragment Size. Finite Case. Lemma 3 although elegant is a limiting result. We check the veracity of the tail distribution of the maximum fragment size (12) for finite n. It turns out to be quite a lot of work. The value of (12) evaluated at $x = (1/k)(\log k + \log \omega)$ is to be compared with Lemma 4.

Lemma 4. *For k sufficiently large, $\mathbb{P}(F_{(1)} \geqslant \frac{n}{k}(\log k + \omega)) = O(e^{-\omega})$.*
If $k \geqslant 2$ is finite, the above becomes $\mathbb{P}(F_{(1)} \geqslant \frac{n}{k-1}(\log k + \omega)) = O(e^{-\omega})$.

Proof. Recall that $N(k) = \binom{n-1}{k-1}$ is the number of partitions of $s = n - k$ unlabelled vertices among k distinguishable root vertices (the influencers). Let $N(\ell, k)$ be the number of these partitions which contain at least one fragment of size ℓ; thus consisting of a root and $\ell - 1$ follower vertices. Using the 'stars and

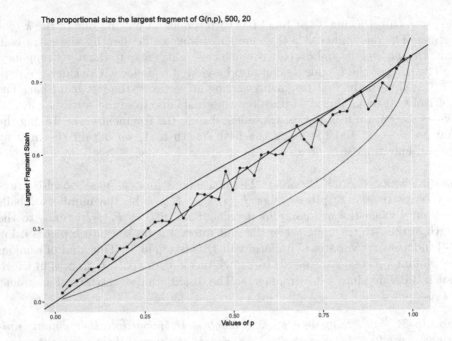

Fig. 3. The blue data plot is the average largest fragment size obtained by simulation using $G(n, p)$. For large $m = Np$, $G(n, p) \sim G(n, m)$ so the results are equivalent. The upper line is Lemma 3 for the largest fragment $S_{(k)}$, where k is based on an estimate of $\mathbb{E}\, a(m)$ using (4). The diagonal black line is $y = x$ for comparison. The lower line is $1/k$, the average component size. The plot is for 20 replications at $n = 500$, values of p interpolated at 0.02. (Color figure online)

bars' notation given above, there are k ways to choose a left hand bar (a cell) to which we allocate $\ell - 1$ stars. There remain $s' = s - (\ell - 1) = n - k - \ell + 1$ stars to be allocated. Contract the specified cell (stars and delimiters) to a single delimiter. The number of delimiters is now $k' = k - 1$, and $n' = s' + k' = n - \ell$. The remaining cells can be filled in $\binom{n'-1}{k'-1}$ ways, and thus

$$N(\ell, k) = k \binom{n - \ell - 1}{k - 2}.$$

Assume $k \geqslant 2$ so that $\ell \leqslant n - 1$. Let $P_k(\ell)$ be the proportion of partitions which contain *at least one fragment* of size ℓ. Then

$$
\begin{aligned}
P_k(\ell) &= \frac{N(\ell, k)}{N(k)} = \frac{N(\ell, k)}{\binom{n-1}{k-1}} \\
&= k(k-1) \frac{(n-k)\ldots(n-k-\ell+2)}{(n-1)\cdots(n-\ell+1)(n-\ell)} \\
&= \frac{k(k-1)}{n-\ell} \left(1 - \frac{k-1}{n-1} \right) \cdots \left(1 - \frac{k-1}{n-\ell+1} \right).
\end{aligned}
$$

Case k tends to infinity. Suppose $k \geqslant \omega$. For any value of k, the expected length of a fragment is n/k so

$$\mathbb{P}(\ell \geqslant \sqrt{k}\mathbb{E}\,\ell) \leqslant \frac{1}{\sqrt{k}}.$$

Assume $\ell \leqslant \sqrt{k}\mathbb{E}\,\ell = n/\sqrt{k}$. We continue with the asymptotics of $P_k(\ell)$.

$$P_k(\ell) = \frac{k(k-1)}{n-\ell} \exp\left(-(k-1)\left(\frac{1}{n-1} + \cdots + \frac{1}{n-\ell+1}\right)\right)$$

$$\times \exp\left(-(k-1)O\left(\frac{1}{n^2} + \cdots + \frac{1}{(n-\ell)^2}\right)\right)$$

$$\sim \frac{k(k-1)}{n-\ell} \exp\left(-(k-1)\log\frac{n-1}{n-\ell}\right) \exp\left(-O(k\ell/n^2)\right)$$

$$= \frac{k(k-1)}{n-\ell}\left(\frac{n-\ell}{n-1}\right)^{k-1} e^{-O(\sqrt{k}/n)} \sim k^2\frac{n-1}{(n-\ell)^2}\left(1 - \frac{\ell+1}{n-1}\right)^k$$

$$\sim \frac{k^2}{n}\exp\left(-\frac{k(\ell+1)}{n-1}\right)e^{-O(k(\ell/n)^2)} \sim \frac{k^2}{n}\exp\left(-\frac{k\ell}{n}\right).$$

Let

$$P_k(\ell \geqslant \ell_0) = P_k(\ell_0 \leqslant \ell < n/\sqrt{k}) + P_k(\ell \geqslant n/\sqrt{k}) = \sum_{\ell=\ell_0}^{n/\sqrt{k}} P_k(\ell) + O(1/\sqrt{k}).$$

Then

$$P(\ell \geqslant \ell_0) \sim \frac{k^2}{n}\sum_{\ell=\ell_0}^{n/\sqrt{k}} e^{-\ell\frac{k}{n}} \sim e^{-\ell_0 k/n}\,\frac{k^2}{n}\,\frac{1 - e^{-O(\sqrt{k})}}{1 - e^{-k/n}} \sim k e^{-\ell_0 k/n}.$$

Let $\ell_0 = (n/k)(\log k + \omega)$, where $\ell_0 \ll n/\sqrt{k}$ then

$$P_k\left(\ell \geqslant \frac{n}{k}(\log k + \omega)\right) \sim e^{-\omega} + O(1/\sqrt{k}). \qquad (13)$$

Thus segments of order $(n/k)(\log k)$ exists with constant probability provided ω is constant. This gives the order of the maximum segment length.

Case: $k \geqslant 2$ finite or tending slowly to infinity. Returning to a previous expression

$$P_k(\ell) \sim \frac{k(k-1)}{n-\ell}\left(\frac{n-\ell}{n-1}\right)^{k-1}.$$

Put $\ell/n = x$ then returning to the expansion of $P_k(\ell)$, and assuming $k \geqslant 3$,

$$P_k(\ell \geqslant \ell_0) \sim \frac{k(k-1)}{n-1}\sum_{\ell \geqslant \ell_0}\left(\frac{n-\ell}{n-1}\right)^{k-2}$$

$$\sim \frac{k(k-1)}{n-1}\int_{\ell_0/n}^{1}(1-x)^{k-2}dx = k(1 - \ell_0/n)^{k-1}.$$

This is similar to the previous case.

References

1. Axelrod, R.: The dissemination of culture: a model with local convergence and global polarization. J. Conflict Resolut. **41**(2), 203–226 (1997)
2. Bennett, P., Dudek, A.: A gentle introduction to the differential equation method and dynamic concentration. Discrete Math. **345**(12), 113071 (2022)
3. Feller, W.: An Introduction to Probability Theory and its Applications. Volume I
4. Flache, A., Mäs, M., Feliciani, T., Chattoe-Brown, E., Deffuant, G., Huet, S., Lorenz, J.: Models of social influence: towards the next frontiers. JASSS **20**(4) 2 (2017). http://jasss.soc.surrey.ac.uk/20/4/2.html
5. Freedman, D.A.: On tail probabilities for martingales. Ann. Probab. **3**, 100–118 (1975)
6. Holst, L.: On the lengths of the pieces of a stick broken at random. J. Appl. Prob. **17**, 623–634 (1980)
7. Moussaïd, M, Kä mmer, J.E., Analytis, P.P., Neth, H.: Social influence and the collective dynamics of opinion formation. PLoS ONE **8**(11), e78433 (2013)
8. Mukhopadhyay, A., Mazumdar, R.R., Roy, R.: Voter and majority dynamics with biased and stubborn agents. J. Stat. Phys. **181**(4), 1239–1265 (2020). https://doi.org/10.1007/s10955-020-02625-w
9. Pyke, R.: Spacings. JRSS(B) **27**(3), 395–449 (1965)
10. Pymar, R., Rivera, N.: On the stationary distribution of the noisy voter model. arXiv:2112.01478 (2021)
11. Yildiz, E., Ozdaglar, A., Acemoglu, D., Saberi, A., Scaglione, A.: Binary opinion dynamics with stubborn agents. ACM Trans. Econ. Comput. **1**(4), 19:1–19:30 (2013)

The Iterated Local Transitivity Model for Tournaments

Anthony Bonato[✉] and Ketan Chaudhary

Toronto Metropolitan University, Toronto, Canada
abonato@torontomu.ca

Abstract. A key generative principle within social and other complex networks is transitivity, where friends of friends are more likely friends. We propose a new model for highly dense complex networks based on transitivity, called the Iterated Local Transitivity Tournament (or ILTT) model. In ILTT and a dual version of the model, we iteratively apply the principle of transitivity to form new tournaments. The resulting models generate tournaments with small average distances as observed in real-world complex networks. We explore properties of small subtournaments or motifs in the ILTT model and study its graph-theoretic properties, such as Hamilton cycles, spectral properties, and domination numbers. We finish with a set of open problems and the next steps for the ILTT model.

1 Introduction

The vast volume of data mined from the web and other networks from the physical, natural, and social sciences, suggests a view of many real-world networks as self-organizing phenomena satisfying common properties. Such complex networks capture interactions in many phenomena, ranging from friendship ties in Facebook, to Bitcoin transactions, to interactions between proteins in living cells. Complex networks evolve via several mechanisms such as preferential attachment or copying that predict how links between nodes are formed over time. Key empirically observed properties of complex networks such as the small world property (which predicts small distances between typical pairs of nodes and high local clustering) have been successfully captured by models such as preferential attachment [2,3]. See the book [4] for a survey of early complex network models, along with [12].

Balance theory cites mechanisms to complete triads (that is, subgraphs consisting of three nodes) in social and other complex networks [17,19]. A central mechanism in balance theory is *transitivity*: if x is a friend of y, and y is a friend of z, then x is a friend of z; see, for example, [22]. Directed networks of ratings or trust scores and models for their propagation were first considered in [18]. *Status theory* for directed networks, first introduced in [21], was motivated by both trust propagation and balance theory. While balance theory focuses on likes and

Research supported by a grant from NSERC.

M. Dewar et al. (Eds.): WAW 2023, LNCS 13894, pp. 179–192, 2023.
https://doi.org/10.1007/978-3-031-32296-9_12

dislikes, status theory posits that a directed link indicates that the creator of the link views the recipient as having higher status. For example, on Twitter or other social media, a directed link captures one user following another, and the person they follow may be of higher social status. Evidence for status theory was found in directed networks derived from Epinions, Slashdot, and Wikipedia [21]. For other applications of status theory and directed triads in social networks, see also [20,24].

The *Iterated Local Transitivity* (*ILT*) model introduced in [8,9] and further studied in [5,10,23], simulates structural properties in complex networks emerging from transitivity. Transitivity gives rise to the notion of *cloning*, where a new node x is adjacent to all of the neighbors of some existing node y. Note that in the ILT model, the nodes have local influence within their neighbor sets. The ILT model simulates many properties of social networks. For example, as shown in [9], graphs generated by the model densify over time and exhibit bad spectral expansion. In addition, the ILT model generates graphs with the small-world property, which requires graphs to have low diameter and high clustering coefficient compared to random graphs with the same number of nodes and expected average degree. A directed analogue of the ILT model was introduced in [6].

Tournaments are directed graphs where each pair of nodes shares exactly one directed edge (or arc). Tournaments are simplified representations of highly interconnected structures in networks. For example, we may consider users on Twitter or TikTok that follow each other or those on Reddit focused on a particular topic or community. A tournament arises from such social networks by assigning an arc (u, v) if the user u responds more frequently to the posts of v than v does to u. Other examples of tournaments in real-world networks include those in sports, with edges corresponding to one player or team winning over another. Such directed cliques are of interest in network science as one type of *motif*, which are certain small-order significant subgraphs. Tournaments in social and other networks grow organically; therefore, it is natural to consider models simulating their evolution.

The present paper explores a version of the ILT model for tournaments. The *Iterated Local Transitivity for Tournaments (ILTT)* model is deterministically defined over discrete time-steps as follows. The only parameter of this deterministic model is the initial tournament $G = G_0$, which is called the *base*. For a non-negative integer t, G_t represents the tournament at time-step t. Suppose that the tournament G_t has been defined for a fixed time-step $t \geq 0$. To form G_{t+1}, for each $x \in V(G_t)$, add a new node x' called the *clone* of x. We refer to x as the *parent* of x', and x' as the *child* of x. We add the arc (x', x), and for each arc (x, y) in G_t, we add the arc (x', y') to G_{t+1}. For each arc (x, y) in G_t, we add arcs (x', y) and (x, y') in G_{t+1}. See Fig. 1. We refer to G_t as an *ILTT tournament*. Note that the subtournament of clones in G_{t+1} is isomorphic to G_t; further, clones share the same adjacencies as their parents.

The *dual* of a tournament reverses the orientation of each of its directed edges. The analogously defined Dual Iterated Local Transitivity for Tournaments (or $ILTT_d$) shares all arcs as defined in the ILTT model, but replaces the subtour-

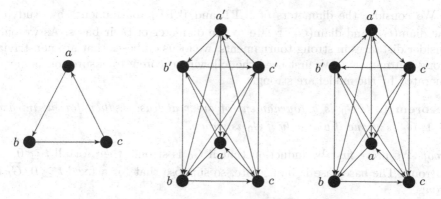

Fig. 1. If G_0 is the directed 3-cycle, then $\overrightarrow{G_1}$ is in the middle and $\overleftarrow{G_1}$ is on the right.

nament of clones by its dual: if (x, y) is an arc in G_t, then (y', x') is an arc in G_{t+1}. To better distinguish the models, we use the notation $\overrightarrow{G_t}$ and $\overleftarrow{G_t}$ for an ILTT and ILTT_d tournament at time-step $t \geq 1$, respectively. See Fig. 1.

We will demonstrate that the ILTT and ILTT_d models simulate properties observed in complex graphs and digraphs, such as the small-world property. Section 2 considers distances in ILTT and ILTT_d tournaments and shows that their average distances are bounded above by a constant. We consider motifs in Sect. 3 and show that while the family of ILTT_d tournaments are universal in the sense that they eventually contain any given tournament (as proved in Theorem 5), the same does not hold for ILTT tournaments. Section 4 focuses on graph-theoretic properties of the models, such as Hamiltonian and spectral properties, and their domination number. We conclude with future directions and open problems.

Throughout the paper, we consider finite, simple tournaments. A tournament is *strong* if for each pair of nodes x and y, there are directed paths connecting x to y and y to x. If $x_1, x_2, \ldots, x_n$ are nodes of a tournament G so that (x_i, x_j) is an arc whenever $i < j$, then G is a *linear order*. A linear order on three nodes is a *transitive 3-cycle*. For background on tournaments, the reader is directed to [1] and to [26] for background on graphs. For background on social and complex networks, see [4, 15].

2   Small World Property

Tournaments are highly dense structures, so local clustering or densification properties are less interesting. However, the presence of short distances remains relevant. In a strong tournament G, the distance from nodes x to y is the length of the shortest path from x to y, denoted $d_G(x, y)$, or $d(x, y)$ if G is clear from context. Although d_G satisfies the triangle inequality, it is not a metric as it need not be symmetric; that is, d_G is a quasimetric. The maximum distance among pairs of nodes in G is its *diameter*, written $\text{diam}(G)$.

We consider the diameters of ILTT and $ILTT_d$ tournaments by studying how $\text{diam}(\overrightarrow{G_t})$ and $\text{diam}(\overleftarrow{G_t})$ relate to the diameter of their bases. As we only consider distances in strong tournaments, we focus on bases that are non-trivial strong tournaments. We first need the following theorem to ensure tournaments generated by the model are strong.

Theorem 1. *If G_0 is a tournament of order at least 3, then for all integers $t \geq 1$, $\overrightarrow{G_t}$ is strong if and only if G_0 is strong.*

Proof. We prove first by induction that if G_0 is strong, then for all $t \geq 0$, G_t is strong. The base case is immediate, so suppose that for a fixed $t \geq 0$, $\overrightarrow{G_t}$ is strong.

The subtournament of $\overrightarrow{G_{t+1}}$ induced by the clones is strong. Let u and v be distinct nodes of $\overrightarrow{G_t}$, so both of these nodes are in the same strong component. We show that u and v' are in the same strong component. Let P be a directed path from u to v in $\overrightarrow{G_t}$, and let (w, v) be the final arc of P. Let P_1 be the directed path obtained by replacing (w, v) with (w, v') in P and let P_2 be a directed path from v' to u'.

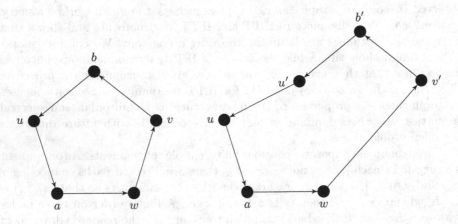

Fig. 2. A cycle passing through u and v in G (left) can be used to obtain a directed cycle passing through u and v' in $\overrightarrow{G}$ (right).

If we traverse along P_1 from u to v', traverse P_2, and then follow the arc (u', u), we have a directed cycle containing u and v'. See Fig. 2 for an example. Note that this also proves that u and u' are in the same strong component. Hence, each pair of nodes of $\overrightarrow{G_{t+1}}$ are in the same strong component, and the induction follows.

For a fixed $t \geq 0$, if $\overrightarrow{G_t}$ is not strong, then there are nodes x and y of $\overrightarrow{G_t}$ so there is no directed path from x to y. If there was a directed path P from x to y in $\overrightarrow{G_{t+1}}$, then by exchanging each clone for its parent in P, we would find a directed path from x to y in $\overrightarrow{G_t}$, which is a contraction. Hence, $\overrightarrow{G_{t+1}}$ is not strong.   □

We consider an analogous result on strong ILTT$_d$ tournaments.

Theorem 2. *Suppose that G_0 has no sink. We then have that for all $t \geq 1$, $\overleftarrow{G_t}$ is strong.*

Proof. We prove the case that $\overleftarrow{G_1}$ is strong, with the cases $t \geq 2$ following in an analogous way by induction. For a node x in G_0, there is some y such that (x, y) is an arc. The directed cycle with arcs $(x, y'), (y', x'), (x', x)$ shows that x and x' are in the same strong component of $\overleftarrow{G_1}$.

Suppose w_1 and w_2 are two nodes in G. Without loss of generality, suppose that (w_1, w_2) is an arc. As G_0 has no sink, there is some node u such that (w_2, u) is an arc. The arcs

$$(w_1, w_2), (w_2, u'), (u', w_2'), (w_2', w_1'), (w_1', w_1)$$

form a directed cycle containing w_1, w_2, w_1', and w_2'. In particular, any two nodes in G_0, any two clones, and any node of G_0 along with a clone are in the same strong component. Hence, $\overleftarrow{G_1}$ is strong.   □

Observe that if G_0 has a sink, then $\overleftarrow{G_1}$ may not be strong, as is the case for G_0 equaling a single directed edge.

We next prove the following lemma on distances in ILTT.

Lemma 1. *Suppose that G_0 is strong with order at least 3. For all integers $t \geq 1$ and for distinct nodes x and y of $\overrightarrow{G_t}$, we have that*

$$d_{\overrightarrow{G_t}}(x, y) = d_{\overrightarrow{G_{t+1}}}(x, y) = d_{\overrightarrow{G_{t+1}}}(x, y') = d_{\overrightarrow{G_{t+1}}}(x', y) = d_{\overrightarrow{G_{t+1}}}(x', y').$$

Proof. Let x and y be nodes of $\overrightarrow{G_t}$. If $\alpha \in \{x, x'\}$ and $\beta \in \{y, y'\}$, then for any directed path P from x to y, replacing x with α in the first arc of P and replacing y with β in the final arc of P is a directed path from α to β in $\overrightarrow{G_{t+1}}$. Thus, we have that $d_{\overrightarrow{G_{t+1}}}(\alpha, \beta) \leq d_{\overrightarrow{G_t}}(x, y)$.

Let $f : V(\overrightarrow{G_{t+1}}) \to V(\overrightarrow{G_t})$ be the function mapping each clone to its parent and fixing each parent. We then have that f is a homomorphism from $\overrightarrow{G_{t+1}}$ to the tournament obtained from $\overrightarrow{G_t}$ by attaching a loop to each node. Let $\alpha \in \{x, x'\}$ and $\beta \in \{y, y'\}$ as in the first paragraph of the proof. Omitting loops from the image of any path from α to β under f, we obtain a directed walk from x to y in G. Thus, $d_{\overrightarrow{G_t}}(x, y) \leq d_{\overrightarrow{G_{t+1}}}(\alpha, \beta)$. The proof follows.   □

In an ILTT tournament with a strong base, for each node x in $\overrightarrow{G_t}$, we have that $d(x, x') = 3$. To see this, s $\overrightarrow{G_t}$ is strong, x is in a directed 3-cycle C. Suppose that C has nodes x, u, and v and arcs (x, u), (u, v), and (v, x). We then have that $(x, u'), (u', v'), (v, x'),$ and (x', x) is a directed cycle in $\overrightarrow{G_1}$. It follows that x and x'

are in the same strong component. We note that there can be no directed 2-path from x to x', and so $d(x,x') = 3$. Also, note that by the definition of the model, $d(x',x) = 1$. Hence, we have the following corollary from these observations and by Lemma 1.

Corollary 1. *If $G = G_0$ is a strong tournament of order at least 3, then an ILTT tournament G_t satisfies* $\mathrm{diam}(\overrightarrow{G_t}) \leq \max\{\mathrm{diam}(G_0), 3\}$.

It is straightforward to see that $d(x,x') = 2$ in $\overleftarrow{G_t}$. By a result analogous to Lemma 1 for distances in the ILTT_d model (with proof omitted), we may also derive that if G_0 is strong and order at least 3, then $\mathrm{diam}(\overleftarrow{G_t}) \leq \mathrm{diam}(G_0)$.

We next consider the average distance of tournaments from the ILTT and ILTT_d models. The *Wiener index* of a strong tournament G of order n is

$$W(G) = \sum_{(u,v) \in V(G) \times V(G)} d_G(u,v).$$

The *average distance* of G is

$$L(G) = \frac{W(G)}{n(n-1)}.$$

Note that we do not have to divide by $\binom{n}{2}$ as distances are not necessarily symmetric.

The following formula for $W(\overrightarrow{G_t})$ follows from Lemma 1 and the definitions.

Theorem 3. *Suppose that G_0 is strong of order $n \geq 3$. If $t \geq 1$ is an integer, then*

$$W(\overrightarrow{G_{t+1}}) = 4(2^t n + W(\overrightarrow{G_t})).$$

Theorem 3 allows us to prove the following corollary, which shows that ILTT tournaments satisfy the small world property. Note that ILTT tournaments tend not to deviate much in average length from that of the base when the base has large order.

Corollary 2. *Let G_0 be a strong tournament of order $n \geq 3$. We have the following properties.*

1. $W(\overrightarrow{G_t}) = 2^{t+1}(2^t - 1)n + 4^t W(G_0)$.
2. $L(\overrightarrow{G_t}) \sim \frac{2 + (n-1)L(G_0)}{n}$.

Proof. The proof follows by induction on $t \geq 0$ with the base case immediate. For the inductive step, by Theorem 3 and the inductive hypothesis, we derive that

$$W(\overrightarrow{G_{t+1}}) = 4 \cdot 2^t n + 4W(\overrightarrow{G_t})$$
$$= n\left(\sum_{k=0}^{t} 2^k 4^{t+1-k}\right) + 4^{t+1}W(G_0).$$

As the geometric series $\sum_{k=0}^{t} 2^k 4^{t+1-k}$ equals $2^{t+2}(2^{t+1}-1)$, item (1) follows.

For item (2), we have by item (1) that

$$
\begin{aligned}
L(G_t) &= \frac{W(\overrightarrow{G_t})}{2^t n(2^t n - 1)} \\
&= \frac{2^{t+1}(2^t - 1)n}{2^t n(2^t n - 1)} + \frac{4^t W(G_0)}{2^t n(2^t n - 1)} \\
&= 2\frac{2^t - 1}{2^t n - 1} + (n-1)L(G_0)\frac{2^t}{2^t n - 1} \\
&\sim \frac{2}{n} + \frac{(n-1)L(G_0)}{n}.
\end{aligned}
$$

The proof follows. □

Interestingly, the average distances in ILTT$_d$ tournaments approach the same limit independent of the base. The proof of the following lemma is omitted for space considerations.

Lemma 2. *Let G_0 be a strong tournament on $n \geq 3$ nodes, and let α be the number of arcs in G not on a directed 3-cycle. For all $t \geq 1$, we then have that*

$$
W(\overleftarrow{G_t}) = 12\binom{2^{t-1}n}{2} + \alpha + 3 \cdot 2^{t-1}n.
$$

The following corollary gives the average distance for ILTT$_d$ tournaments.

Corollary 3. *Let G_0 be a strong tournament of order at least 3. We then have that*

$$
\lim_{t \to \infty} L(\overleftarrow{G_t}) = \frac{3}{2}.
$$

Proof. Let α be the number of arcs not on a directed 3-cycle in G_0. By Lemma 2 we have that for $t \geq 1$

$$
\begin{aligned}
L(\overleftarrow{G_t}) &= \frac{W(\overleftarrow{G_t})}{2^t n(2^t n - 1)} \\
&= \frac{6(2^{t-1}n(2^{t-1}n - 1)) + \alpha + 3(2^{t-1}n)}{2^t n(2^t n - 1)} \\
&\sim \frac{3}{2}.
\end{aligned}
$$

The proof follows. □

3   Motifs and Universality

Motifs are small subgraphs that are important in complex networks as one measure of similarity. For example, the counts of 3- and 4-node subgraphs give a similarity measure for distinct graphs; see [7] for implementations of this approach using machine learning. In the present section, we give results on subtournaments in ILTT and ILTT_d.

The following theorem shows that every linear order is a subtournament of an ILTT or an ILTT_d tournament.

Theorem 4. *For an integer* $t \geq 1$, *the linear order of order* t *is a subtournament of* $\overrightarrow{G_t}$ *and* $\overleftarrow{G_t}$.

Proof. We give the proof for the ILTT model as the argument is analogous to the ILTT_d model. The proof follows by induction on $t \geq 1$. Suppose $\{x_1, x_2, \ldots, x_t\}$ is set of nodes of a linear order in $\overrightarrow{G_t}$. Let $x_0 = x_1'$ in $\overrightarrow{G_{t+1}}$. We then have that the subtournament on $\{x_0, x_1, x_2, \ldots, x_t\}$ is a linear order with $t+1$ nodes in $\overrightarrow{G_{t+1}}$.   □

Despite Theorem 4, the subtournaments of ILTT tournaments are restricted by their base tournament. As an application of Theorem 1 (or seen directly), note that if we begin with a transitive 3-cycle as G_0, then for all $t \geq 0$, $\overrightarrow{G_t}$ never contains a directed 3-cycle.

Given a tournament on some finite set of nodes, one can obtain any other tournament on those nodes by simply reversing the orientations of certain arcs. Given two tournaments G and H on the same set of nodes and $m \geq 1$ an integer, we say that G and H *differ by* m *arcs* if there are m distinct pairs of nodes x and y such that (x, y) is an arc of G while (y, x) is an arc of H. We have the following lemma.

Lemma 3. *Let* G_0 *be a tournament. For* $t \geq 0$, *if* H *is a subtournament of* $\overleftarrow{G_t}$ *and* H *and a tournament* J *differ by one arc, then* J *is isomorphic to a subtournament* $\overleftarrow{G_{t+1}}$.

Proof. Let (u, v) be an arc of H such that (v, u) is an arc of J. We replace u by u' and v by v' in $\overleftarrow{G_{t+1}}$. The subtournament on $(V(H) \setminus \{u, v\}) \cup \{u', v'\}$ is an isomorphic copy of J in $\overleftarrow{G_{t+1}}$. See Fig. 3.   □

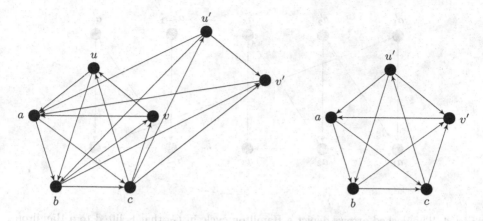

Fig. 3. Constructing an isomorphic copy of J in $\overleftarrow{G_{t+1}}$, which results by reversing the arc (v, u).

Inductively extending Lemma 3 to tournaments differing by more than one arc and combining this fact with Theorem 4, we find that *any* tournament is a subtournament of an $ILTT_d$ tournament with an arbitrary base. We say that the family of $ILTT_d$ tournaments are *universal*. Note that we do not attempt to optimize the value of κ_n in the statement of the theorem.

Theorem 5. *For each $n \geq 2$, let $\kappa_n = n + \binom{n}{2}$. A tournament J of order n nodes is isomorphic to a subtournament of some $\overleftarrow{G_r}$, where $r \leq \kappa_n$.*

Proof. By Theorem 4, we may find an isomorphic copy of the linear order L_n with n nodes in $\overleftarrow{G_n}$. We may now iteratively apply Lemma 3 to L_n, reversing arrows as needed until we arrive at an isomorphic copy of J in some $\overleftarrow{G_r}$. As J and L_n differ by at most $\binom{n}{2}$ arcs, we have that $r \leq \kappa_n$. $\square$

4   Graph-Theoretic Properties of the Models

This section presents various graph-theoretic results on the ILTT and $ILTT_d$ models. We consider the Hamilitonicity, spectral properties, and the domination number of tournaments generated by the model.

4.1   Hamiltonicity

A *Hamiltonian cycle* in a tournament G is a directed cycle that visits each node of G exactly once. The tournament G is said to be *Hamiltonian* if it has a Hamilton cycle. Suppose C_1 is a Hamilton cycle in a tournament G of order $r > 3$ passing through nodes $a_1, a_2, \ldots, a_r$, and a_1 in that order. Consider the cycle C_2 passing through the nodes a_k, a'_{k+1}, and a_{k+1} in that order for k ranging over $\{1, 2, \ldots, r\}$ and subscripts taken modulo r; See Fig. 4.

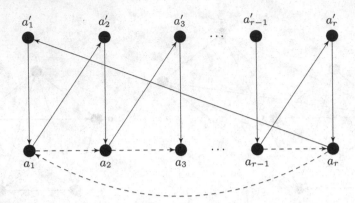

Fig. 4. The dashed arrows depict a Hamilton cycle in G_0 that is lifted to a Hamilton cycle in $\overrightarrow{G_1}$ (or $\overleftarrow{G_1}$), depicted by the solid arrows.

In both $\overrightarrow{G_1}$ and $\overleftarrow{G_1}$, C_2 is a Hamilton cycle. In particular, in C_2 each a_k has a unique in-neighbor a'_{k-1} and out-neighbor a_{k+1}, and each a'_k has in-neighbor a_{k-1} and out-neighbor a_k. Therefore, the following theorem follows by induction on $t \geq 1$.

Theorem 6. *If G_0 is a Hamiltonian tournament of order at least three, then for all $t \geq 1$, so are $\overrightarrow{G_t}$ and $\overleftarrow{G_t}$.*

Camion [13] proved that if G is a tournament of order at least three, then G is Hamiltonian if and only if G is strong. We, therefore, have the following corollary of Theorem 6.

Corollary 4. *Suppose that the base tournament G_0 is of order at least 3.*

1. *An ILTT tournament is Hamiltonian if and only if its base is strong.*
2. *An $ILTT_d$ tournament is Hamiltonian if its base has no sink.*

4.2 Spectral Properties

We next consider eigenvalues of the adjacency matrices of ILTT tournaments. Spectral graph theory is a well-developed area for undirected graphs (see [14]) but less so for directed graphs (where the eigenvalues may be complex numbers with non-zero imaginary parts). The following theorem characterizes the eigenvalues of ILTT tournaments.

Theorem 7 below gives a characterization of all non-zero eigenvalues of the tournaments arising from the ILTT model in terms of the non-zero eigenvalues of the base. Figure 5 gives a visualization of the eigenvalues in an example of an ILTT tournament.

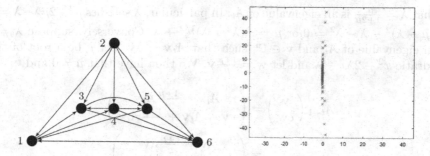

Fig. 5. If G_0 is the tournament depicted on the left, the right figure depicts the eigenvalues of the ILTT tournament $\overrightarrow{G_7}$ in the complex plane.

Theorem 7. *If A_t is the adjacency matrix of an ILTT tournament $\overrightarrow{G_t}$ of order n, then the following items hold.*

1. *The value $-1/2$ is not an eigenvalue of A_{t+1}.*
2. *For each non-zero eigenvalue λ of A_t, $\mu = \pm(\lambda^2 + \lambda)^{1/2} + \lambda$ is an eigenvalue of A_{t+1} and these are all the non-zero eigenvalues of A_{t+1}.*

Proof. Suppose $\mu \neq 0$ is an eigenvalue of A_{t+1} and let $\mathbf{v}, \mathbf{w} \in \mathbb{C}^n$ not both the zero vector $\mathbf{0}$ be such that $\mu \begin{pmatrix} \mathbf{v} \\ e\mathbf{w} \end{pmatrix} = A_{t+1} \begin{pmatrix} \mathbf{v} \\ e\mathbf{w} \end{pmatrix}$. We then have that $A_{t+1} = \begin{pmatrix} A_t & A_t \\ I + A_t & A_t \end{pmatrix}$. Hence,

$$\mu\mathbf{v} = A_t\mathbf{v} + A_t\mathbf{w} \tag{1}$$
$$\mu\mathbf{w} = \mathbf{v} + A_t\mathbf{v} + A_t\mathbf{w}. \tag{2}$$

We then have that $\mu\mathbf{w} = (1 + \mu)\mathbf{v}$. By (1), it follows that $\frac{1+\mu}{\mu}A_t\mathbf{v} + A_t\mathbf{v} = \mu\mathbf{v}$, which gives that

$$\frac{1 + 2\mu}{\mu}A_t\mathbf{v} = \mu\mathbf{v}.$$

For (1), note that, by (1) and (2), if $\mu = -1/2$, then

$$A_t\mathbf{v} + A_t\mathbf{w} = -\frac{1}{2}\mathbf{v}$$
$$\mathbf{v} + A_t\mathbf{v} + A_t\mathbf{w} = -\frac{1}{2}\mathbf{w}.$$

Thus, $\mathbf{v} = -\mathbf{w}$ and $\mathbf{v} = \mathbf{0}$, which is a contradiction. Therefore, $\mu \neq -1/2$.

For (2), note that the computation in the proof of part (1) also shows that

$$A_t\mathbf{v} = \frac{\mu^2}{1 + 2\mu}\mathbf{v},$$

so that $\lambda = \frac{\mu^2}{1+2\mu}$ is an eigenvalue of A_t. In particular, λ satisfies $\mu^2 - 2\mu\lambda - \lambda = 0$ or $(\mu - \lambda)^2 - \lambda - \lambda^2 = 0$ or $\mu = \pm(\lambda^2 + \lambda)^{1/2} + \lambda$. Conversely, suppose $\lambda \neq 0$ is an eigenvalue of A_t and $\mathbf{v} \in \mathbb{C}^n$ such that $A_t\mathbf{v} = \lambda\mathbf{v}$. Let μ be a root of the quadratic $x^2 - 2\lambda x - \lambda$ and let $\mathbf{w} = \frac{1+\mu}{\mu}\mathbf{v}$. We then have that $\mu \neq 0$ and

$$A_{t+1}\begin{pmatrix} \mathbf{v} \\ e\mathbf{w} \end{pmatrix} = \begin{pmatrix} A_t\mathbf{v} + \frac{1+\mu}{\mu}A_t\mathbf{v} \\ e\mathbf{v} + A_t\mathbf{v} + \frac{1+\mu}{\mu}A_t\mathbf{v} \end{pmatrix}$$

$$= \begin{pmatrix} \lambda\mathbf{v} + \frac{1+\mu}{\mu}\lambda\mathbf{v} \\ e(1+\lambda)\mathbf{v} + \frac{1+\mu}{\mu}\lambda\mathbf{v} \end{pmatrix}$$

$$= \begin{pmatrix} (2\lambda + \frac{\lambda}{\mu})\mathbf{v} \\ e(1+2\lambda + \frac{\lambda}{\mu})\mathbf{v} \end{pmatrix}.$$

The equation $\mu^2 - 2\lambda\mu - \lambda = 0$ implies that $\frac{\lambda}{\mu} = \mu - 2\lambda$. Hence,

$$A_{t+1}\begin{pmatrix} \mathbf{v} \\ e\mathbf{w} \end{pmatrix} = \begin{pmatrix} \mu\mathbf{v} \\ e(1+\mu)\mathbf{v} \end{pmatrix} = \mu\begin{pmatrix} \mathbf{v} \\ e\mathbf{w} \end{pmatrix},$$

so that μ is an eigenvalue of A_{t+1}. □

4.3 Domination Numbers

Dominating sets contain nodes with strong influence over the rest of the network. Several earlier studies focused on the domination number of complex network models; see [11,16]. In a tournament G, an *in-dominating set* S has the property that for each u not in S, there is a $v \in S$ such that (u, v). The dual notion is an *out-dominating set*. The cardinality of a minimum order in-dominating set is the *in-domination number* of G, written $\gamma^-(G)$; the dual notion is the *out-domination number* of G, written $\gamma^+(G)$.

The following theorem describes the evolution of the domination numbers of ILTT and ILTT$_d$ tournaments. Its proof is omitted due to space considerations.

Theorem 8. *If G_0 is a tournament, then the following identities hold for all $t \geq 1$:*

1. $\gamma^\pm(G_0) = \gamma^\pm(\overrightarrow{G_t})$.
2. $\gamma^+(G_0) = \gamma^+(\overleftarrow{G_t})$.
3. $\gamma^-(G_0) \leq \gamma^-(\overleftarrow{G_t})$.

Note that a consequence of Theorem 8 is that the domination numbers of ILTT tournaments equal the domination numbers of their base. An analogous fact holds for the ILTT$_d$ model in the case of the out-domination number.

5   Conclusion and Further Directions

We considered new models ILTT and ILTT$_d$ for evolving, complex tournaments based on local transitivity. We proved that both models generate tournaments with small distances and that ILTT$_d$ tournaments contain all tournaments as subtournaments. We investigated when tournaments from the models were strong or Hamiltonian, and considered their in- and out-domination numbers. We also considered spectral properties of ILTT tournaments.

Studying the counts of various motifs in the models, such as directed and transitive cycles, would be interesting. Insights into the motifs found in the model may shed light on the resulting tournament limits (which are an extension of graphons to tournaments; see [25]). We did not consider spectral properties of ILTT$_d$ tournaments, and the cop number and automorphism group of tournaments generated by the model would be interesting to analyze. We may also consider randomized versions of the models, where in each time-step, either the subtournament of clones is either the existing tournament or its dual, chosen with a given probability. It would also be interesting to consider the effects of placing a random tournament on the subtournament of clones, where arcs are oriented with a given probability.

References

1. Bang-Jensen, J., Gutin, G.: Digraphs. Springer, London (2009). https://doi.org/10.1007/978-1-84800-998-1
2. Barabási, A., Albert, R.: Emergence of scaling in random networks. Science **286**, 509–512 (1999)
3. Bollobás, B., Riordan, O., Spencer, J., Tusnády, G.: The degree sequence of a scale-free random graph process. Random Struct. Algorithms **18**, 279–290 (2001)
4. Bonato, A.: A Course on the Web Graph. American Mathematical Society Graduate Studies Series in Mathematics, Providence, Rhode Island (2008)
5. Bonato, A., Chuangpishit, H., English, S., Kay, B., Meger, E.: The iterated local model for social networks. Discret. Appl. Math. **284**, 555–571 (2020)
6. Bonato, A., Cranston, D.W., Huggan, M.A., Marbach, T.G., Mutharasan, R.: The Iterated Local Directed Transitivity model for social networks. In: Proceedings of WAW 2020 (2020)
7. Bonato, A., et al.: Dimensionality matching of social networks using motifs and eigenvalues. PLoS ONE **9**(9), e106052 (2014)
8. Bonato, A., Hadi, N., Prałat, P., Wang, C.: Dynamic models of on-line social networks. In: Proceedings of WAW 2009 (2009)
9. Bonato, A., Hadi, N., Horn, P., Prałat, P., Wang, C.: Models of on-line social networks. Internet Math. **6**, 285–313 (2011)
10. Bonato, A., Infeld, E., Pokhrel, H., Prałat, P.: Common adversaries form alliances: modelling complex networks via anti-transitivity. In: Proceedings of WAW 2017 (2017)
11. Bonato, A., Lozier, M., Mitsche, D., Perez Gimenez, X., Prałat, P.: The domination number of online social networks and random geometric graphs. In: Proceedings of the 12th Conference on Theory and Applications of Models of Computation (2015)

12. Bonato, A., Tian, Y.: Complex networks and social networks. In: Kranakis, E. (ed.) Advances in Network Analysis and its Applications. Mathematics in Industry, vol. 18, pp. 269–286. Springer, Heidelberg (2011). https://doi.org/10.1007/978-3-642-30904-5_12

13. Camion, P.: Chemins et circuits hamiltoniens des graphes complets. Comptes Rendus de l'Académie des Sciences de Paris **249**, 2151–2152 (1959)

14. Chung, F.R.K.: Spectral Graph Theory. American Mathematical Society, Providence, Rhode Island (1997)

15. Chung, F.R.K., Lu, L.: Complex Graphs and Networks. American Mathematical Society, USA (2004)

16. Cooper, C., Klasing, R., Zito, M.: Lower bounds and algorithms for dominating sets in web graphs. Internet Math. **2**, 275–300 (2005)

17. Easley, D., Kleinberg, J.: Networks, Crowds, and Markets Reasoning about a Highly Connected World. Cambridge University Press, Cambridge (2010)

18. Guha, R.V., Kumar, R., Raghavan, P., Tomkins, A.: Propagation of trust and distrust. In: Proceedings of WWW (2004)

19. Heider, F.: The Psychology of Interpersonal Relations. Wiley, Hoboken (1958)

20. Huang, J., Shen, H., Hou, L., Cheng, X.: Signed graph attention networks. In: Proceedings of Artificial Neural Networks and Machine Learning - ICANN (2019)

21. Leskovec, J., Huttenlocher, D.P., Kleinberg, J.M.: Signed networks in social media. In: Proceedings of the ACM SIGCHI (2010)

22. Scott, J.P.: Social Network Analysis: A Handbook. Sage Publications Ltd., London (2000)

23. Small, L., Mason, O.: Information diffusion on the iterated local transitivity model of online social networks. Discret. Appl. Math. **161**, 1338–1344 (2013)

24. Song, D., Meyer, D.A.: A model of consistent node types in signed directed social networks. In: Proceedings of the 2014 IEEE/ACM International Conference on Advances in Social Networks Analysis and Mining (2014)

25. Thörnblad, E.: Decomposition of tournament limits. Eur. J. Comb. **67**, 96–125 (2018)

26. West, D.B.: Introduction to Graph Theory, 2nd edn. Prentice Hall, Hoboken (2001)

Author Index

A

Alaluusua, Kalle 83
Avrachenkov, Konstantin 83

B

Banerjee, Sayan 147
Betlen, Andrei 36
Bonato, Anthony 179
Brandes, Ulrik 99

C

Chaudhary, Ketan 179
Chung, Fan 112
Cooper, Colin 164

D

Dehghan, Ashkan 36
Deka, Prabhanka 147
Dvořák, Michal 68

G

Gösgens, Martijn 1
Gracar, Peter 19

H

Hasheminezhad, Rouzbeh 99

J

Joslyn, Cliff 127

K

Kamiński, Bogumił 36, 52
Kang, Nan 164
Kay, Bill 127
Knop, Dušan 68
Kumar, B. R. Vinay 83

L

Leskelä, Lasse 83
Litvak, Nelly 1
Lüchtrath, Lukas 19

M

Miller, David 36
Misiorek, Paweł 52
Mönch, Christian 19
Myers, Audun 127

O

Olvera-Cravioto, Mariana 147

P

Prałat, Paweł 36, 52
Purvine, Emilie 127

R

Radzik, Tomasz 164
Roek, Gregory 127
Rønberg, August Bøgh 99

S

Schierreich, Šimon 68
Shapiro, Madelyn 127
Sieger, Nicholas 112
Siuta, Kinga 36
Skorupka, Agata 36

T

Théberge, François 52

V

van der Hofstad, Remco 1

M. Dewar et al. (Eds.): WAW 2023, LNCS 13894, p. 193, 2023.
https://doi.org/10.1007/978-3-031-32296-9

Printed in the United States
by Baker & Taylor Publisher Services